Y0-ACA-971

Communication in Development

Communication in Development

The Twenty-Eighth Symposium
The Society for Developmental Biology

Boulder, Colorado, June 16–18, 1969

Communication in Development

Organized and Edited by

Anton Lang

MSU/AEC Plant Research Laboratory
Michigan State University
East Lansing, Michigan 48823

Developmental Biology, Supplement 3

Editor-in-Chief

M. V. Edds, Jr.

1969

ACADEMIC PRESS, New York and London

ACADEMIC PRESS, INC.
111 Fifth Avenue, New York, New York 10003

United Kingdom Edition published by
ACADEMIC PRESS, INC. (LONDON) LTD.
Berkeley Square House, London W.1

LIBRARY OF CONGRESS CATALOG CARD NUMBER: 55-10678

PRINTED IN THE UNITED STATES OF AMERICA

Communication In Development

Contributors and Presiding Chairmen

Numbers in parentheses indicate the pages on which the authors' contributions begin.

First Session: How It All Came About

Chairman: E. Margoliash, Scientific Divisions, Abbott Laboratories, North Chicago, Illinois 60064

H. H. PATTEE, W. W. Hansen Laboratories of Physics, Stanford University, Stanford, California 94305 (1)

J. E. VARNER, MSU/AEC Plant Research Laboratory, Michigan State University, East Lansing, Michigan 48823 (17)

Second Session: Much Later—The Cell Communing with Itself, Part I.

Chairman: D. M. Prescott, Department of Molecular, Cellular and Developmental Biology, University of Colorado, Boulder, Colorado 50302

OLE MAALØE, Institute of Microbiology, University of Copenhagen, Øster Farimagsgade 2A, Copenhagen, Denmark (33)

J. B. GURDON, Department of Zoology, Oxford University, Parks Road, Oxford, England (59)

Second Session, Part II.

Chairman: C. P. Wolk, MSU/AEC Plant Research Laboratory, Michigan State University, East Lansing, Michigan 48823

LIONEL F. JAFFE, Department of Biological Sciences, Purdue University, Lafayette, Indiana 47907 (83)

Third Session: Cells Talking with Cells

Chairman: P. Grant, Department of Biology, University of Oregon, Eugene, Oregon 97403

FRANK L. ADLER AND MARVIN FISHMAN, Department of Immunology, The Public Health Research Institute of the City of New York, Inc., 455 First Avenue, New York, New York 10016 (112)

Fourth Session: The Organism Conversing with Its Cells, Part I.

Chairman: H. Schneiderman, Developmental Biology Center, Case Western Reserve University, Cleveland, Ohio 44106

CARROLL M. WILLIAMS, Department of Biology, Harvard University, 16 Divinity Avenue, Cambridge, Massachusetts 02138 (133)

Fourth Session, Part II.

Chairman: R. Hertel, MSU/AEC Plant Research Laboratory, Michigan State University, East Lansing, Michigan 48823

E. V. JENSEN, M. NUMATA, S. SMITH, T. SUZUKI, P. I. BRECHER, AND E. R. DESOMBRE, The Ben May Laboratory for Cancer Research, University of Chicago, 950 East 59th Street, Chicago, Illinois 60637 (151)

PETER M. RAY, Department of Biological Sciences, Stanford University, Stanford, California 94305 (172)

Fifth Session: The Environment Instructing the Cell, Part I.

Chairman: R. T. Schimke, Department of Pharmacology, Stanford University Medical Center, Stanford, California 94305

PHILIP FILNER, MSU/AEC Plant Research Laboratory, Michigan State University, East Lansing, Michigan 48823 (206)

Fifth Session, Part II.

Chairman: W. S. Hillman, Department of Biology, Brookhaven National Laboratory, Upton, New York 11973

STERLING B. HENDRICKS, Plant Industry Station, U. S. Department of Agriculture, Beltsville, Maryland 20705 (227)

Contents

II. Much Later—The Cell Communing with Itself

III. Cells Talking with Cells

V. The Environment Instructing the Cell

DEVELOPMENTAL BIOLOGY SUPPLEMENT 3, 1–16 (1969)

How Does a Molecule Become a Message?

H. H. PATTEE

W. W. Hansen Laboratories of Physics, Stanford University, Stanford, California 94305

INTRODUCTION

The theme of this symposium is "Communication in Development," and, as an outsider to the field of developmental biology, I am going to begin by asking a question: How do we tell when there *is* communication in living systems? Most workers in the field probably do not worry too much about defining the idea of communication since so many concrete, experimental questions about developmental control do not depend on what communication means. But I am interested in the origin of life, and I am convinced that the problem of the origin of life cannot even be formulated without a better understanding of how molecules can function symbolically, that is, as records, codes, and signals. Or as I imply in my title, to understand origins, we need to know how a molecule becomes a message.

More specifically, as a physicist, I want to know how to distinguish *communication* between molecules from the normal physical *interactions* or forces between molecules which we believe account for all their motions. Furthermore, I need to make this distinction *at the simplest possible level*, since it does not answer the origin question to look at highly evolved organisms in which communication processes are reasonably clear and distinct. Therefore I need to know how messages originated.

Most biologists will say that, while this is an interesting question, there are many problems to be solved about "how life works," before we worry about how it all began. I am not going to suggest that most of the "how it works" problems have been solved, but at the same time I do not see that knowing much more about "how it works" in the current style of molecular biology and genetics is likely to lead to an answer to origin problems. Nothing I have learned from molecular biology tells me in terms of basic physical principles why matter should ever come alive or why it should evolve along an entirely different pathway than inanimate matter. Furthermore, at every hierarchical level of biological organization we are presented with very much the same kind of problem. Every evolutionary innovation amounts to a new level of integrated control. To see how this integrated control works, that is, to see how the physical implementation of this control is accomplished, is not the same as understanding how it came to exist.

The incredible successes of biochemistry in unraveling the genetic code and the basic mechanism of protein synthesis may suggest that we can proceed to the next hierarchical level with assurance that if we pay enough attention to molecular details, then all the data will somehow fall into place. I, for one, am not at all satisfied that this kind of answer even at the level of replication should be promulgated as the "secret of life" or the "reduction of life to ordinary physics and chemistry," although I have no doubt that some of these molecular descriptions are a necessary step toward the answer. I am even less satisfied that developmental programs will be comprehended only by taking more and more molecular data.

Let me make it quite clear at this point that I believe that all the molecules in the living cell obey precisely the laws of normal physics and chemistry (Pattee, 1969). We are not trying to understand molecular structure, but language structure in the most elementary sense, and this means understanding not only "how it works," but how it originated. Nor do I agree with Polanyi's (1968) conclusion that the constraints of language and machines are "irreducible"; although I do believe Polanyi has presented this problem—a problem which is too often evaded by molecular biologists—with the maximum clarity. Whatever the case may be, it is not likely that an acceptable resolution of either origin or reduction problems will come about only by taking more data. I believe we need both a theory of the origin of hierarchical organization as well as experiments or demonstrations showing that the hierarchical constraints of a "language" can actually originate from the normal physical constraints that hold molecules together and the laws which govern their motions.

It is essential in discussions of origins to distinguish the sequence of causal events from the sequence of control events. For example, the replicative controls of cells harness the molecules of the environment to produce more cells, and the developmental controls harness the cells to produce the organism; so we can say that development is one level higher than replication in the biological hierarchy. One might argue then that insofar as developmental messages turn off or on selected genes in single cells according to specific interactions with neighboring cells, they can only be a later evolutionary elaboration of the basic rules of self-replication.

However, I believe we must be very cautious in accepting the conclusion of the evolutionary sequence too generally, and especially in extending it to the origin of life. Single, isolated cells clearly exhibit developmental controls in the growth of their structure, so that messages must be generated by interactions of the growing cell with its own structure, so to speak. But since this characteristic structure is certainly a part of the "self" which is being replicated, it becomes unclear how to separate the developmental from the replicative controls. Furthermore, it is one of the most general characteristics of biological evolution that life has increasingly buffered itself from the changes and ambient conditions of the environments. This buffering is accomplished by establishing hierarchical levels of control that grow more and more distinct in their structure and function as evolution progresses. But we must remember that these hierarchical levels always become blurred at their origin. Therefore, when viewing a highly evolved hierarchical organization we must not confuse the existing control chains in the final hierarchical system with the causal chains or evolutionary sequence of their origin.

Our own symbolic languages have many examples of hierarchical structure which do not correspond to a causal order or the sequence in which the structures appeared (e.g., Lenneburg, 1967). The evolution of all hierarchical rules is a bootstrap process. The rules do not create a function—they improve an existing function. The functions do not create the rules—they give the rules meaning. For example, stoplights do not account for how people drive—they help people drive more effectively. Nor does traffic create stop lights—traffic is the reason why stop lights make sense.

Therefore it is reasonable to consider the hypothesis that the first "messages" were expressed not in the highly integrated and precise genetic code that we find today, but in a more global set of geophysical and geochemical constraints, which we could call the primeval "ecosystem language," from which the genetic code condensed in much the same way that our formal rules of syntax and dictionaries condensed from the functional usage of primitive symbols in a complex environment. If this were indeed the case, then it would be more likely that "developmental replication" in the form of external cycles not only preceded autonomous "self-replication," but may have accounted for the form of the genetic code itself.

SOME PROPERTIES OF LANGUAGES AND SYMBOLS

The origin of languages and messages is inseparable from the origin of arbitrary rules. It is a general property of languages and symbol systems that their constraints are arbitrary in the sense that the same function can be accomplished by many different physical and logical structures. For example in the case of human language we find many symbol vehicles and alphabets, many dictionaries and syntactical rules, and many styles of writing, all of which function adequately for human communication. The same is true for the machine languages which man has invented to communicate with computers; and as for the physical embodiment of these language structures it is clear, at least in the case of the machine, that the particular physical structures which perform the logic, memory, reading and writing functions are almost incidental and have very little to do with the essential logical constraints of the language system itself.

The arbitrariness in primitive biological languages is less clear. We know that there are many examples of differing organ design with essentially the same function. On the other hand, the universality of the genetic code could be used as an argument against arbitrariness in biological languages. This would be a weak argument at present, however, since the origin of the code is completely unknown. Furthermore, the only experimental evidence, which is meager, indirectly supports the "frozen accident" theory (Crick, 1968) which implies that almost any other code would also work.

The "frozen accident" theory also illustrates what I have found to be a principle of hierarchical structures in general, a principle that may be stated as a principle of impotence: Hierarchical organizations obscure their own origins as they evolve. There are several ways to interpret this. We may think of a hierarchical control as a collective constraint or rule imposed on the motion of individual elements of the collection. For such a constraint to appear as a "rule" it must be much simpler than the detailed motions of the elements. The better the hierarchical rule, the more selective it is in measuring particular details of the elements it is constraining. For example, a good stoplight system does not measure all the dynamical details of the traffic, but only the minimum amount of information about the time and direction of cars which, in principle at least, makes the traffic flow as safely and rapidly as practical. This essential simplification, or loss of detail is also what obscures the origin of the rule.

This ill-defined property of simplification is common to all language and machine constraints, and hierarchical systems in general —that the essential function of the system is "obscured" by too many details of how it works. One well-known example is our spoken language. If while speaking about these problems I were to begin thinking about the details of what I am saying—the syntax of my sentences, my pronunciation, how the symbols will appear on the printed page—I would rapidly lose the function of communication, which was the purpose of all these complex constraints of the language in the first place. In the same way the function of a computer, or for that matter an automobile or a watch, would be lost if to use them we always had to analyze the mechanical details of their components. I would say that the secret of good communication in general lies in knowing what to ignore rather than in finding out in great detail what is going on.

Therefore as a preliminary answer to our first question of how we distinguish communication between molecules from the normal physical interactions, I suggest that one necessary condition for the appearance of a message is that very *complex interactions lead to a very simple result*. The nonliving world, at least as viewed by the physicist, often ends up the other way, with the simplest possible problem producing a very complicated result. The more details or degrees of freedom that the physicist considers in his problem the more complex and intricate becomes the solution. This complexity grows so rapidly with the number of particles that the physicist very quickly resorts to a drastic program of relinquishing *all* detailed knowledge, and then talks only about the statistics of very large aggregations of particles. It is only through some "postulate of ignorance" of the dynamical details that these statistical descriptions can be used consistently. Even so, the passage from the dynamical description to the statistical description in physics poses very deep problems which are unavoidably related to the communication of information or messages from the physical system to the observer (Brillouin, 1962). If we accept this general idea that communication is in some way a simplification of a complex dynamical process, then we are led by the origin problem to consider what the simplest communication system can be. Only by conceiving of a language in the most elementary terms can we hope to distinguish what is really essential from the "frozen accidents."

WHAT IS THE SIMPLEST MESSAGE?

The biological literature today is full of words like activator, inhibitor, repressor, derepressor, inducer, initiator, regulator. These general words describe *messengers*, specific examples of which are being discovered every day. I would simplify the messages in all these cases by saying they mean "turn on" or "turn off." It is difficult to think of a simpler message. But taken by itself, outside the cell or the context of some language, "turn on" is not really a message since it means nothing unless we know from where the signal came and what is turned on as a result of its transmission. It is also clear that the idea of sending and receiving messages involves a definite time sequence and a collection of alternative messages. "Turn on" makes no sense unless it is related by a temporal as well as by a spatial network. On the other hand, one must not be misled by the apparent simplicity of this message. For when such simple messages are concatenated in networks, logicians have shown us that the descriptive potential of such "sequential switching machines" or "automata" are incredibly rich, and that in a formal sense they can duplicate many of the most complex biological activities including many aspects of thought itself. Almost all molecular biological systems operate in this discrete, on-off mode rather than by a continuous modulation type of control. Since many essential input and output variables are continuous, such as concentration gradients and muscle movements, this poses the serious problem, familiar to logicians as well as computer designers, of transcribing discrete variables into continuous variables and vice versa. The transcription process also determines to a large degree the simplicity as well as the reliability of the function.

If the simplest message is to turn something on, then we also need to know the physical origin and limits of the simplest device that will accomplish this operation. Such a device is commonly called a *switch*, and we shall use this term, bearing in mind that it is defined by its function, not by our design of artificial switches that we use to turn on lights or direct trains. The switch is a good example of an element with an exceedingly simple function—it is hard to imagine a simpler function—but with a detailed behavior, expressed in terms of physical equations of motion, which is exceedingly complex. Switches in certain forms, such as ratchets and Maxwell demons, have caused physicists a great deal of difficulty. In a way, this is contrary to our intuition since even a small child can look at a switch or a ratchet and tell us "how it works." With considerably more effort, using more sophisticated physical and chemical techniques, it may soon be possible to look at allosteric enzyme switches and explain "how they work."

We must bear in mind, however, that in both cases there are always deeper levels of answers. For example, the physical description "how it works" is possible only if we ignore certain details of the dynamical motion. This is because the switching event which produces a single choice from at least two alternatives is not symmetrical in time and must therefore involve dissipation of energy, that is, loss of detailed information about the motions of the particles in the switch. As a consequence of this dissipation or loss of detail it is physically impossible for a switch to operate with absolute precision. In other words, no matter how well it is designed or how well it is built, all devices operating as switches have a finite probability of being "off" when they should be "on," and vice versa. This is not to say that some switches are not better than others. In fact the enzyme switches of the cell have such high speed and reliability compared with the artificial switches made by man that it is doubtful if their behavior can be explained quantitatively in terms of classical models. Since no one has yet explained a switch in terms of quantum mechanics, the speed and reliability of enzymes remains a serious problem for the physicist (Pattee, 1968). But even though we cannot yet explain molecular switches in terms of fundamental physics, we can proceed here by simply assuming their existence and consider under what conditions a network of switches might be expected to function in the context of a language.

WHAT IS THE SIMPLEST NATURAL LANGUAGE?

We come now to the crucial question. An isolated switch in nature, even if we could explain its origin, would have no function in the sense that we commonly use the word. We see here merely the simplest possible instance of what is perhaps the most fundamental problem in biology—the question of how large a system one must consider before biological *function* has meaning. Classical biology generally considers the cell to be the minimum unit of life. But if we consider life as distinguished from nonliving matter by its evolutionary behavior in the course of time, then it is clear that the isolated cell is too small a system, since it is only through the communication of cells with the outside environment that natural selection can take place. The same may be said of developmental systems in which collections of cells create messages that control the replication and expression of individual cells.

The problem of the origin of life raises this same question. How large a system must we consider in order to give meaning to the idea of life? Most people who study the origin of life have made the assumption that the hierarchical structure of highly evolved life tells us by its sequence of control which molecules came first on the primeval earth. Thus, it is generally assumed that some form of non-enzymatic, self-replicating nucleic acid first appeared in the sterile ocean, and that by random search some kind of meaningful message was eventually spelled out in the sequence of bases, though it is never clear from these descriptions how this lonely "message" would be read. Alternatively, there are some who believe the first important molecules were the enzymes or the switches which controlled metabolic processes in primitive cell-like units. I find it more reasonable to begin, not with switching mechanisms or meaningless messages, but rather with a primitive communication *network* which could be called the primeval ecosystem. Such a system might consist of primitive geochemical matter cycles in which matter is catalytically shunted through cell-like structures which occur spontaneously without initial genetic instructions or metabolic control. In my picture, it is the constraints of the primeval ecosystem which, in effect, generate the language in which the first specific messages can make evolutionary sense. The course of evolution by natural selection will now produce better, more precise, messages as measured in this ecological language; and in this case signals from the outside world would have preceded the autonomous genetic controls which now originate inside the cell.

But these speculations are not my main point. What I want to say is that *a molecule does not become a message because of any particular shape or structure or behavior of the molecule. A molecule becomes a message only in the context of a larger system of physical constraints which I have called a "language"* in analogy to our normal usage of the concept of message. The trouble with this analogy is that our human languages are far too complex and depend too strongly on the structure and evolution of the brain and the whole human organism to clarify the problem. We are explaining the most simple language in terms of the most complex. Anyway, since the origin of language is so mysterious that linguists have practically given up on the problem, we cannot expect any help even from this questionable analogy. What approaches, then, can we find to clarify what we mean by the simplest message or the simplest language?

THE SIMPLEST ARTIFICIAL LANGUAGES

The most valuable and stimulating ideas I have found for studying the origin of language constraints has come from the logicians and mathematicians, who also try to find the simplest possible formal languages which nevertheless can generate an infinitely rich body of theorems. A practical aspect of this problem is to build a computer with the smallest number of switches which can give you answers to the maximum number of problems. This subject is often called "automata theory" or "computability theory," but it has its roots in symbolic logic, which is itself a mathematical language to study all mathematical languages. This is why it is of such interest to mathematicians: all types of mathematics can be developed using this very general language. The basic processes of replication, development, cognitive activity, and even evolution, offer an intriguing challenge to the automata theorist as fundamental conceptual and logical problems, and also to the computer scientist who now has the capability of "experimental" study of these simulated biological events. There is often a considerable communication gap between the experimental biologist and the mathematician interested in biological functions, and this is most unfortunate, for it is unlikely that any other type of problem requires such a comprehensive approach to achieve solutions.

But let us return to our particular problem of the origin of language structure and messages. What can we learn from studying artificial languages? As I see it, the basic difficulty with computer simulation is that whenever we try to invent a model of an elementary or essential biological function, the program of our model turns out to be unexpectedly complex if it actually accomplishes the defined function in a realistic way. The most instructive examples of this that I know are the models of self-replication. I shall not discuss any of these in detail, but only give the "results." It is possible to imagine many primitive types of mechanical, chemical, and logical processes which perform some kind of replication (e.g., Penrose, 1958; Pattee, 1961; Moore, 1962). It is also quite obvious that most of these systems have no conceivable evolutionary potential, nor can one easily add on any developmental elaborations without redesigning the whole system or causing its failure.

The first profound model of a self-replicating system that I know, was that of the mathematician John von Neumann (1956), who explicitly required of his model that it be capable of evolving a more elaborate model without altering its basic rules. von Neumann was influenced strongly by the work of Turing (1937), who carried the concept of computation to the simplest extreme in terms of basic operations with symbols, and showed that with these basic rules one can construct a "universal" machine which could compute any function that any other machine could compute. von Neumann also made use of the McCulloch and Pitts (1943) models of neuronal switching networks in his thinking about replication, but he extended both these models to include a "construction" process, which was not physically realistic, but which allowed him to describe a "universal self-replicating automaton" which had the potential for evolution and to which developmental programs could be added without changing the basic organization of the automaton.

But what was the significance of such a model? What impressed von Neumann was the final complexity of what started out as the "simplest" self-replicating machine that could evolve. He concluded that there must be a "threshold of complexity" necessary to evolve even greater complexity, but below which order deteriorates. Furthermore, this threshold appeared to be so complex that its spontaneous origin was inconceivable.

Since von Neumann's work on self-replication, there have been further serious logical attempts to simplify or restate the problem (e.g., Arbib, 1967a; Thatcher, 1963). Automata theory has also been used to describe developmental processes (e.g., Apter and Wolpert, 1965; Arbib, 1967b). But the basic results are the same. If the program does anything which could be called interesting from a biological point of view, or if it can even be expected to actually work as a program on any real computer, then such programs turn out to be unexpectedly complex with no hint as to how they could have originated spontaneously. For example, one of the simplest models of morphogenesis is the French Flag problem, in which it is required that a sheet of self-replicating cells develop into the pattern of the French Flag. This can be done in several ways (e.g., Wolpert, 1968), but the program is not nearly as simple as one might expect from the simplicity of the final pattern it produces.

It is the common feeling among automata theorists, as well as computer programmers, that if one has never produced a working, developmental, replicative, or evolutionary program, then one is in for a discouraging surprise. To help popularize this fact, Michie and Longuet-Higgins (1966) published a short paper called "A Party Game Model of Biological Replication" which will give some idea of the logic to the reader who has had no computer experience. But as computer scientists emphasize, there is no substitute for writing a program and making it work.

Why are all biological functions so difficult to model? Why is it so difficult to imitate something which looks so simple? Indeed, functional simplicity is not easy to achieve, and very often the more stringent the requirements for simplicity of function, the more difficult will be the integration of the dynamical details necessary to carry out the function. While it is relatively easy to imagine *ad hoc* "thought machines" that will perform well-defined functions, the structure of real machines is always evolved through the challenges of the environment to what are initially very poorly defined functions. These challenges usually have more to do with how the machine fails than how it works. In other words, it is the *reliability*, *stability*, or *persistence* of the function, rather than the abstract concept of the pure function itself, which is the source of structure. We can see this by studying the evolution of any of our manmade machines. Of course in this case man himself defines the general function, but how the structure of the machine finally turns out is not determined by man alone. The history of timepieces is a good example. It is relatively easy to see superficially with each escapement or gear train "how it works," but only by understanding the requirements of precision and stability for "survival," as well as the environmental challenges to these requirements in the form of temperature variations, external accelerations, corrosion, and wear, can we begin to understand the particular designs of escapements, gear teeth, and power trains which have survived.

Our understanding of the genetic code and of developmental programs is still at the "how does it work" level, and although we may be able to trace the evolutionary changes, even with molecular detail, we have almost no feeling for which details are crucial and which are incidental to the integrated structure of the organism. The analytical style of molecular biology, which has brought us to this level, first recognizes a highly evolved function and then proceeds to look at the structures in more and more detail until all the parts can be isolated in the test tube, and perhaps reassembled to function again. But if we wish to explain origins or evolutionary innovations, this style may be backward.

If we believe that selective catalysts or "switching molecules" do not make messages by themselves, then we should study not them by themselves, but in switching networks as they might have occurred in a primitive "sterile" ecosystem. Nor should we try, if we are looking for origins, to design switching networks to perform well-defined functions such as universal self-replication or the development of a French Flag morphology, since there is no reason to expect such functions to exist in the beginning. A more realistic approach would be to ask what behavior of more or less random networks of switching catalysts would appear because of its persistence or stability in the face of surrounding disorder. In other words, we should look not for the elements that accomplish well-defined functions, but for the functions that appear spontaneously from collections of well-defined elements. How can this be done?

THE SIMULATION OF ORIGINS

The experimental study of the origin of function or any evolutionary innovation is exceptionally difficult because, to observe such innovation naturally, we must let nature take its course. For the crucial innovations we are discussing, like the origin of molecular messages, language constraints, and codes, nature has already taken its course or is going about it too slowly for us to observe. So again we are left with computer simulation of nature, hoping that the underlying dynamics of the origin of hierarchical organization is so fundamental that it can be observed even in a properly designed artificial environment.

The essential condition for the study of "natural" origins in artificial machines is that we cannot overdefine the function that we hope will originate spontaneously. In other words, we must let the computer take its own course to some degree. A good example of this strategy has been reported by Kauffman (1969). In this example he constructed a "random network" of "random switches" and then observed the behavior. The switches were random in the sense that one of the 2^{2^k} Boolean functions of the k inputs to each switch was chosen at random. Once chosen, however, both the switch function and the network structure connecting inputs and outputs of the switches were fixed.

The significant results were that for low connectivity, that is, two or three inputs per switch, the network produced cycles of activity that were both short and stable—short compared to the enormous number of states, and stable in the sense that the network returns to the same cycle even if a switch in that cycle is momentarily off when it should be on, or vica versa. Kauffman pictured his network as a very simple model of the genetically controlled enzymatic processes in the single cell; I believe, however, this type of model would more appropriately represent a primeval ecosystem in which initially random sequences in copolymer chains begin to act as selective catalysts for further monomer condensations. With the allowance for the creation of new switching catalysts, we would expect condensation of catalytic sequences produced by the switching cycles, to act very much like a primitive set of language constraints. The copolymer sequences would then represent a "record" of the cycle structure.

In our own group, Conrad (1969) has taken a more realistic view of the physical constraints that are likely to exist on the primitive sterile earth, as well as the competitive interactions and requirements for growth that must exist between replicating organism in a finite, closed matter system. These competitive growth constraints have been programmed into an evolutionary model of a multiniche ecosystem with organisms represented by genetic strings subject to random mutation and corresponding phenotypic strings which interact with the other organisms. Although this program includes much more structure than the Kauffman program, neither the species nor the environmental niches are initially constrained by the program, but they are left to find their own type of stability and persistence. The population dynamics is determined, not by solving differential equations that can only represent hypothetical laws, but by actually counting the individuals in the course of evolution of the program. Such a program to a large extent finds its own structure in its most stable dynamical configuration, which we can observe in the course of its evolution.

These computer programs illustrate one approach to the study of the origin of the language constraints we have been talking about. They are empirical studies of the natural behavior of switching networks which do not have specific functions designed into them. This is the way biological constraints must have evolved. But even so, you will ask whether these computer simulations are not too far removed from the biological structures, the cells, enzymes, and hormones that are the real objects of our studies.

This is true—the computer is quite different from a cell—but this disadvantage for most studies of "how it works" is also the strength of such simulation for origin studies. The crucial point I want to make is that the collective behavior we are studying in these models is not dependent on exactly how the individual switches work or what they are made of. We are not studying how the switches work, but how the network behaves. Only by this method can we hope to find developmental and evolutionary principles that are common to all types of hierarchical organizations. Only by studies of this type can we hope to separate the essential rules from the frozen accidents in living organisms.

THE ROLE OF THEORY IN BIOLOGY

There has always been a great difference in style between the physical and biological sciences, a difference which is reflected most clearly in their different attitudes toward theory. Stated bluntly, physics is a collection of basic theories, whereas biology is a collection of basic facts. Of course this is not only a difference in style but also a difference in subject matter. The significant facts of life are indeed more numerous than the facts of inanimate matter. But physicists still hope that they can understand the nature of life without having to learn *all* the facts.

Many of us who are not directly engaged in studying developmental biology or in experimenting with particular systems of communication in cells look at the proliferation of experimental data in developmental biology, neurobiology, and ecology and wonder how all this will end. Perhaps some of you who try to keep up with the literature wonder the same thing. Living systems are of course much more complicated than formal languages, or present computer programs, since living systems actually construct new molecules on the basis of genetic instruction. But even with a few simple rules and small memories, we know it is possible to write "developmental" programs that lead to incredibly rich and formally unpredictable behavior (e.g., Post, 1943). Therefore in the biological sciences it is not altogether reassuring to find that all our data-handling facilities—our journals, our symposia, our mail, and even our largest, quickest computers—are overburdened with information. The physicist Edward Condon once suggested that the whole scientific endeavor will come to an end because this "data collection" does not converge. Certainly if our knowledge is to be effective in our civilization, we must see to it that our theoretical conceptions are based on the elements of simplicity that we find in all our other integrated biological functions; otherwise our knowledge will not survive.

What we may all hope is that the language constraints at all levels of biological organization are similar to the rules of our formal languages, which are finite and relatively simple even though they are sufficient to generate an infinite number of sentences and meanings. We must remember, at the same time, that the potential variety of programs is indeed infinite, and that we must not consume our experimental talents on this endless variety without careful selection based on hypotheses which must be tested. Of course we shall need more experimental data on specific messenger molecules and how they exercise their developmental controls. But to understand how the molecules became messages, and how they are designed and integrated to perform with such incredible effectiveness, we must also account for the reliability of the controlling molecules as well as the challenges and constraints of the ecosystem which controlled their evolution. This in turn will require a much deeper appreciation of the physics of switches and the logic of networks.

ACKNOWLEDGMENT

This work was supported by the National Science Foundation, Grant GB 6932 of the Biological Oceanography Program in the Division of Biological and Medical Sciences.

REFERENCES

Apter, M. J., and Wolpert, L. (1965). Cybernetics and development. *J. Theoret. Biol.* **8,** 244.

Arbib, M. A. (1967a). Some comments on self-reproducing automata. "Systems and Computer Science" (J. F. Hart and S. Takasu, eds.), p. 42. Univ. of Toronto Press, Toronto, Canada.

Arbib, M. A. (1967b). Automata theory and development: Part I. *J. Theoret. Biol.* **14,** 131.

Brillouin, L. (1962). "Science and Information Theory," 2nd ed., Chapters 20 and 21. Academic Press, New York.

Conrad, M. E. (1969). Computer experiments on the evolution of co-adaptation in a primitive ecosystem. Dissertation, Stanford University.

Crick, F. H. C. (1968). The origin of the genetic code. *J. Mol. Biol.* **38,** 367.

Kauffman, S. A. (1969). Metabolic stability and epigenesis in randomly constructed genetic nets. *J. Theoret. Biol.* **22,** 437.

Lenneburg, E. H. (1967). "Biological Foundations of Language." Wiley, New York.

McCulloch, W. S., and Pitts, W. (1943). A logical calculus of the ideas immanent in nervous activity. *Bull. Math. Biophys.* **5,** 115.

Michie, D., and Longuet-Higgins, C. (1966). A party game model of biological replication. *Nature* **212,** 10.

Moore, E. F. (1962). Machine models of self-reproduction. *Proc. Symp. Appl. Math.*, Vol. 14, Mathematical Problems in the Biological Sciences, American Math. Soc., Providence, R. I., p. 17.

Pattee, H. H. (1961). On the origin of macromolecular sequences. *Biophys. J.* **1,** 683.

Pattee, H. H. (1968). The physical basis of coding and reliability in biological evolution. *In* "Towards a Theoretical Biology" (C. H. Waddington, ed.), Vol. 1, p. 67. Edinburgh Univ. Press, Edinburgh, Scotland.

Pattee, H. H. (1969). Physical problems of heredity and evolution. *In* "Towards a Theoretical Biology" (C. H. Waddington, ed.), Vol. 2, p. 268. Edinburgh Univ. Press, Edinburgh, Scotland.

Penrose, L. S. (1958). The mechanics of self-reproduction. *Ann. Human Genet.* **23,** part I, 59.

Polanyi, M. (1968). Life's irreducible structure. *Science* **160,** 1308.

Post, E. L. (1943). Formal reductions of the general combinational decision problem. *Am. J. Math.* **65,** 197.

Thatcher, J. W. (1963). The construction of the self-describing Turing machine. *Proc. Symp. Math. Theory of Automata;* Vol. 12 of the Microwave Research Institute Symposia Series, Brooklyn Polytechnic Press, p. 165.

Turing, A. M. (1937). On computable numbers, with an application to the Entscheidungs problem. *Proc. London Math. Soc.* Ser. 2, **42,** 230.

von Neumann, J. (1956). The general and logical theory of automata. Reprinted in "The World of Mathematics" (J. R. Newman, ed.), Vol. 4., p. 2070. Simon & Schuster, New York.

Wolpert, L. (1968). The French Flag problem: A contribution to the discussion on pattern development and regulation. *In* "Towards a Theoretical Biology," (C. H. Waddington, ed.), Vol. 1, p. 125. Edinburgh Univ. Press, Edinburgh, Scotland.

DEVELOPMENTAL BIOLOGY SUPPLEMENT **3**, 17–32 (1969)

Evolution of Developmental Communication Systems

J. E. VARNER

MSU/AEC Plant Research Laboratory, Michigan State University, East Lansing, Michigan 48823

PROLOGUE

"Darwin placed at the root of life a primordial germ, from which he conceived that the amazing richness and variety of the life now upon the earth's surface might be deduced. If this hypothesis were true, it would not be final. The human imagination would infallibly look behind the germ and, however hopeless the attempt, would enquire into the history of its genesis" (Tyndall, 1871).

INTRODUCTION

It seems likely that many of the principles that underly development, as we now know it, must have applied during the production and selection of the earliest duplicating units. We should, therefore, begin a study of the evolution of development by looking at those conditions thought to exist at the time of the origin of life.

The age of our galaxy and the age of the universe is thought to be 7×10^9 years, or seven eons (Shklovskii and Sagan, 1968). The age of meteorites and the composition of terrestrial leads suggests that both meteorites and leads were involved in some homogenization event about 4.6 eons ago (Fig. 1). This event closely approximates the time of the origin of our solar system and presumably the time of formation of the earth. A minimum, but not a maximum, age of the earth can be set by the age of the oldest minerals dated. This minimum age is 3.5 eons. Sedimentary rocks, which could not have originated without weathering and a hydrosphere, have a minimum age of 3 eons in South Africa, 2.7 eons in Minnesota, and, in general, less in other places. The compositions of these detrital and chemical sediments indicate that there was little or no free oxygen in the atmosphere earlier than 1.8 to 2 eons ago and that there was little ammonia and methane in the atmosphere from about 3 eons onward. According to the sedimentary record, the early atmospheric gases were water, carbon dioxide, carbon monoxide, nitrogen, hydrochloric acid, sulfur dioxide, and a few other gases in trace amounts (Cloud,

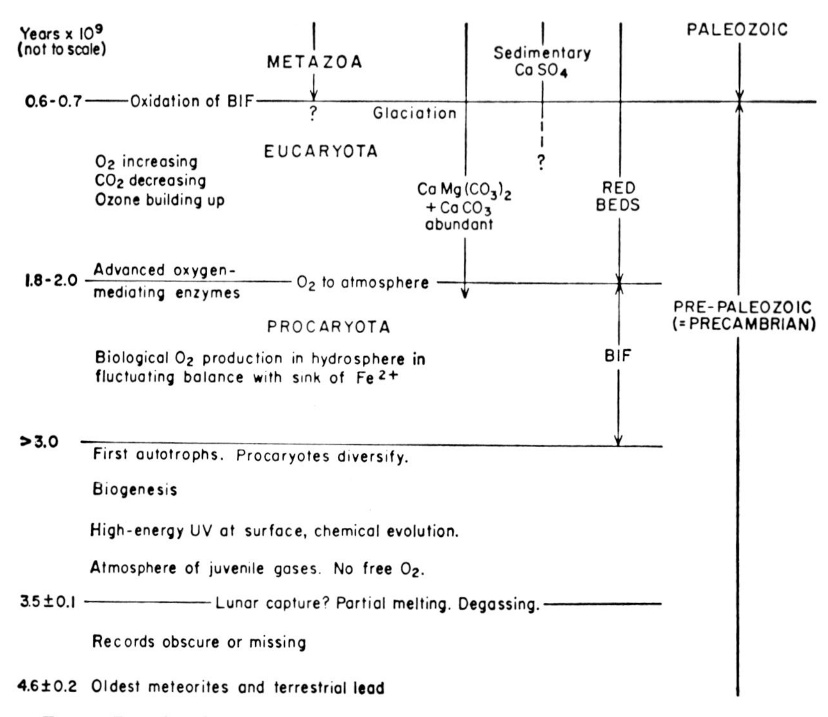

FIG. 1. Postulated main features of interacting biospheric, lithospheric, and atmospheric evolution on the primitive earth. From Cloud (1968).

1968). These gases are thought to have had their origin in the degassing of the earth's interior as a result of melting about 3.5 eons ago. Although it is agreed that the concentration of oxygen in the atmosphere could not have been more than about 0.02%, opinions are divided about the concentration of such gases as hydrogen, methane, and ammonia. It seems likely that in an atmosphere devoid of oxygen and rich in high energy sources such as ultraviolet light and electrical discharges, production of reduced compounds of nitrogen, carbon, and sulfur could occur. Accumulation of these compounds would be favored by entrapment in the primitive oceans. These entrapped compounds would be protected from the harsh effects of ultraviolet light by the water between them and the surface, and from oxidation by the lack of oxygen. We now suppose that in this chemical garden of Eden a chemical evolution occurred as these compounds reacted with themselves, with each other and each others progeny.

Since the early suggestions of Haldane and of Oparin (1968) that life originated in such a hot, dilute soup much experimental progress has been made toward building compounds characteristic of present day life under conditions assumed to simulate prebiotic times (Fox, 1965).

PREBIOLOGICAL MODELS: STATISTICAL

Morowitz (1968; Rider and Morowitz, 1968) has taken the point of view that the examination of prebiological processes is a problem in

TABLE 1

DISTRIBUTION OF BONDS IN 10^{12} WET *Escherichia coli* CELLS AND COMPARABLE DISTRIBUTIONS FOR THE EQUILIBRIUM STATE OF THE SAME ATOMIC COMPOSITION AND FOR THE MOST RANDOM DISTRIBUTION OF BOND ENERGY ISOENERGETIC WITH THE BIOMASS OF *E. coli*[a]

Bond	Equilibrium distribution of bonds $\times 10^{20}$	Distribution of bonds in *E. coli* cell $\times 10^{20}$	Random distribution of bonds isoenergetic to the cell $\times 10^{20}$
Graphite	43[b]	0	0
Ethane	2	0	0
C=C	0	3.00	0
C—C	17	37.94	69
C=N	0	3.00	0
C—N	0	28.75	29
C=O	0	13.37	2
C—O	0	12.25	9
C—H	20	71.03	42
N≡N	6	0	0
N=N	0	0	0
N—N	0	0	3
N—O	0	0	1
N—H	5	17.01	8
O—H	553	505.79	537

[a] From Rider and Morowitz (1968).
[b] Carbon atoms.

$$5\,HCN \longrightarrow (HCN)_5$$

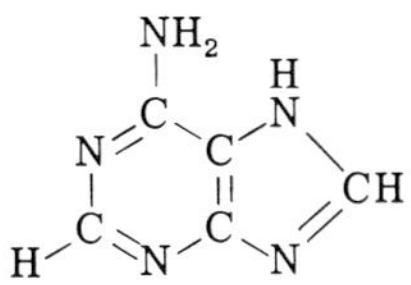

Adenine

FIG. 2. Adenine: a polymer of HCN. After Matthews and Moser (1967)

TABLE 2

PERCENTAGE YIELDS OF pApA OLIGONUCLEOTIDES FORMED ON A POLYURIDYLIC ACID TEMPLATE AFTER 14 DAYS[a]

pH	Polyuridylic acid	pApA	Trimer
6	Without	0.4	—
6	With	4.9	0.2
7	Without	0.7	—
7	With	19.8	2.45
8	Without	1.1	—
8	With	27.6	3.8

[a] After Weimann *et al.* (1968).

nonequilibrium thermal physics. He develops a model system which consists of a box containing carbon, hydrogen, nitrogen, and oxygen atoms placed in contact with an infinite isothermal reservoir at temperature T. Energy from a high potential source flows into the system, and thermal energy flows from the system into the isothermal sink. After some time the system would reach a steady state in which energy flow in would equal heat flow out. Under steady-state conditions, the energy of the system would be above that of the isothermal sink. Assuming that this energy is equally available to all bonds, the most probable covalent bond distribution is then calculated and compared with the actual bond distribution of the cell stuff of *Escherichia coli*. It is clear from Table 1 that the random distribution of bonds in a steady-state system isoenergetic to the cell much more closely resembles the actual distribution of bonds in *E. coli* than does the distribution calculated for equilibrium conditions. This would seem to be a valuable model to use as a starting point in a study of the origin of life.

PREBIOLOGICAL MODELS: CHEMICAL

Beginning with the early and now classic experiments of Miller (1953) in Urey's laboratory in which a mixture of hydrogen, methane, ammonia, and water were subjected to a high voltage electric spark, nearly all classes of biomonomers have been synthesized from these or other simple precursors. It is also interesting that an experiment analogous to Miller's was performed in 1913 by Loeb (reaction 1).

$$CO + NH_3 + H_2O \xrightarrow{\text{spark}} \text{glycine} \tag{1}$$

Also in Calvin's laboratory in 1951 (Garrison *et al.*, 1951), formaldehyde and formic acid were produced from carbon dioxide and water by particles accelerated in the cyclotron (reaction 2).

$$CO + H_2O \xrightarrow{\text{cyclotron}} HCHO + HCOOH \qquad (2)$$

There are now many instances of the formation of polypeptides from amino acids under conditions that might have existed in prebiotic times. For instance, appreciable polymerization of glycine occurs in 2 *N* ammonium hydroxide (Oro and Guidry, 1961) (reaction 3).

$$\text{Glycine} \xrightarrow{2N\ NH_4OH} (\text{glycine})_n \quad n = 2-18 \qquad (3)$$

Mixtures of peptides containing 15 different amino acids are obtained after addition of water to the brown solid which results from the reaction at room temperature of hydrogen cyanide and anhydrous ammonia (Matthews and Moser, 1967). Adenine, $(HCN)_5$ (Fig. 2) is also present in this mixture. Amino acids found in these peptides include lysine, histidine, arginine, aspartic acid, threonine, serine, glutamic acid, glycine, alanine, isoleucine, leucine, valine, glutamine, and asparagine. Heating methane, ammonia, and water over powdered silica produces low yields of all the common nonsulfur

TABLE 3
COMPARISON OF EXPERIMENTALLY DETERMINED DIPEPTIDE YIELDS AND FREQUENCIES CALCULATED FROM KNOWN PROTEIN SEQUENCES[a]

Dipeptide[b]	Values (relative to Gly-Gly)	
	Experimental	Calculated
Gly-Gly	1.0	1.0
Gly-Ala	0.8	0.7
Ala-Gly	0.8	0.6
Ala-Ala	0.7	0.6
Gly-Val	0.5	0.2
Val-Gly	0.5	0.3
Gly-Leu	0.5	0.3
Leu-Gly	0.5	0.2
Gly-Ile	0.3	0.1
Ile-Gly	0.3	0.1
Gly-Phe	0.1	0.1
Phe-Gly	0.1	0.1

[a] From Steinman and Cole (1967).

[b] The dipeptides are listed in terms of increasing volume of the side chains of the constituent residues, Gly = glycine, Ala = alanine, Val = valine, Leu = leucine, Ile = isoleucine, Phe = phenylalanine. Example: Gly-Ala = glycylalanine.

amino acids except histidine and tryptophan (Harada and Fox, 1964). No amino acids of a nonprotein variety were found. Methionine can be produced as shown (Steinman *et al.*, 1968) (reactions 4, 5).

$$NH_3 + CH_4 + H_2O + H_2S \xrightarrow{\text{spark}} NH_4SCN \qquad (4)$$

$$NH_4SCN \xrightarrow[H_2O]{\text{U V light}} \text{methionine} \qquad (5)$$

Porphyrins can be produced by spark discharge in mixtures of methane, ammonia, and water (Hodgson and Ponnamperuma, 1968) and by heating dilute aqueous solutions of pyrrole and formaldehyde (Hodgson and Baker, 1967).

The production of cyanamide and dicyandiamide from hydrogen cyanide is of great interest because these compounds promote in dilute aqueous solutions dehydration condensation reactions (Steinman *et al.*, 1965; Ponnamperuma and Peterson, 1965) (reactions 6 and 7).

$$HCN \xrightarrow[H_2O]{\text{U V light}} \text{dicyandiamide (DCD)}$$

$$\left.\begin{array}{l} P_i + P_i \xrightarrow{\text{DCD}} PP_i \\ \text{Glucose} + P_i \xrightarrow{\text{DCD}} \text{G-6-P} \\ \text{Adenosine} + P_i \xrightarrow{\text{DCD}} \text{AMP} \end{array}\right\} \qquad (6)$$

$$\text{Glycine} + \text{leucine} \xrightarrow[\text{U V light}]{0.01\,M \text{ cyanamide}} \left\{\begin{array}{l} \left.\begin{array}{l}\text{glycyl-glycine} \\ \text{glycyl-leucine} \\ \text{leucyl-glycine} \\ \text{leucyl-leucine}\end{array}\right\} 1\% \text{ yield} \\ \text{glycyl-glycyl-glycine}\} \ 0.1\% \text{ yield} \end{array}\right. \qquad (7)$$

Cyanamide also promotes the formation of deoxyadenosine from adenine and deoxyribose in dilute aqueous solutions (Ponnamperuma and Kirk, 1964) and the formation of uridylic acid from uridine and inorganic phosphate (Lohrman and Orgel, 1968).

Nucleosides and nucleotides can be formed by the ultraviolet irradiation of purine and pyrimidine bases, ribose and ethyl metaphosphate.

Nucleotides also can be produced by the surprisingly easy procedure of heating nucleosides and sodium dihydrogen phosphate to 160°C (Ponnamperuma and Mack, 1965).

Cyanoacetylene, which can be prepared by spark discharge in methane–nitrogen mixtures, adds phosphate in dilute aqueous solutions to form cyanovinylphosphate. Cyanovinylphosphate phosphorylates inorganic phosphate and uridine with excellent yields (Ferris, 1968).

$$H-C \equiv C-CN + HPO_4^{2-} \xrightarrow[\text{pH } 7-9]{H_2O} {}^{=}O_3POCH-CH-CN \text{ (90\% yield)}$$
$$\xrightarrow{P_i} PP_i \text{ (2\%)} \qquad \xrightarrow{\text{uridine}} \text{UMP (4\%)} \tag{8}$$

The selectivity for these phosphorylations is 1.3 for uridine and 9 for inorganic phosphate. That is, uridine and phosphate are phosphorylated in better yield than expected on the basis of their concentration compared with the concentration of water.

Imidazoles are readily formed from simple precursors under potentially prebiotic conditions (Ferris *et al.*, 1968), and activated phosphates react with imidazoles in aqueous solution to give *N*-phosphorimidazole derivatives; thus the prebiotic occurrence of such derivatives is not implausible. Adenosine 5′-monophosphorimidazo-

TABLE 4

Composition of Hydrolyzate of Proteinoid from Amino Acid Adenylates Alone Compared with an Average Protein (Calculated without Ammonia)[a]

Amino acid	Composition of polymer (mole %)	Ratios of amino acids in average protein (mole %)
Lysine	6.5	5.9
Histidine	2.4	1.8
Arginine	4.2	4.9
Aspartic acid	10.3	9.7
Threonine	4.9	4.8
Serine	4.2	6.0
Glutamic acid	9.7	12.7
Proline	5.1	6.2
Glycine	11.1	12.6
Alanine	14.3	9.6
Valine	7.3	5.9
Methionine	0.7	1.8
Isoleucine	4.5	6.0
Leucine	9.6	6.0
Tyrosine	0.1	2.3
Phenylalanine	4.5	3.7

[a] From Vegotsky and Fox (1962).

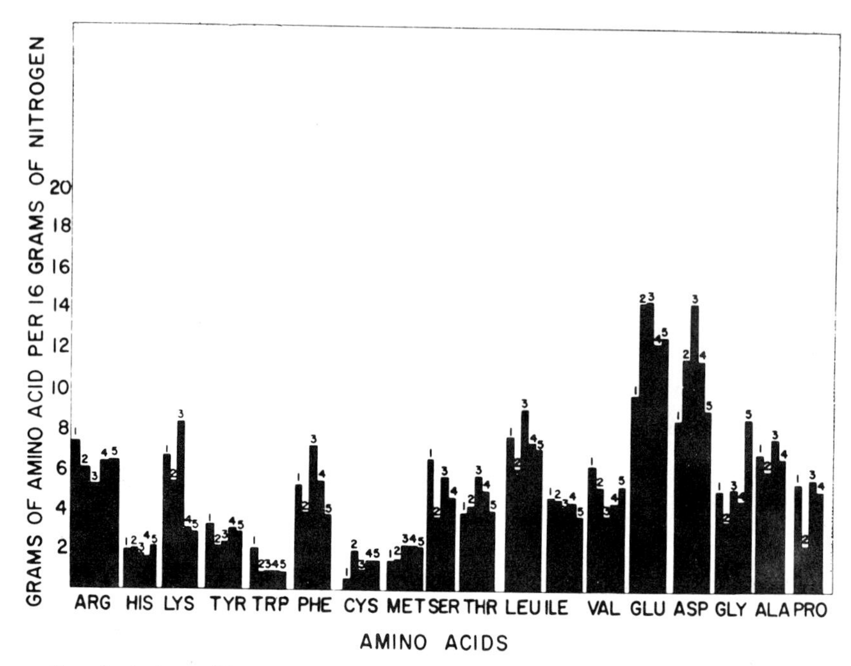

FIG. 3. Amino acid compositions in proteins of different organisms: (1) algae; (2) bacteria; (3) protozoa; (4) invertebrates; (5) mammals. From Vegotsky and Fox (1962).

lide reacts efficiently with adenosine derivatives on a polyuridylic acid template with the formation of internucleotide bonds (Weimann *et al.*, 1968; see Table 2).

The association of amino acids in aqueous solutions under suggested prebiological conditions is not random. The interaction between two amino acids can depend not only on the relative abundance of the amino acids available but also on their pK values and the size of the side chains of the residues involved. That this may have biological relevance is shown by the following. When the yields of dipeptides formed at room temperature from 0.01 M amino acid mixtures in 0.10 N HCl with addition of 0.10 M dicyanamide as a condensing agent are compared with the frequency of occurrence of these same dipeptides calculated from the known sequences of egg lysozyme, ribonuclease, sheep insulin, whale myoglobin, yeast cytochrome c, tobacco mosaic virus, β-corticotropin, glucagon, melanocyte-stimulating-hormone, and chymotrypsin a striking parallel is seen (Steinman and Cole, 1967; see Table 3). Because of the possibility that in early biology peptide chains with reactive side chains could have arisen from nonpolar members of an already formed polymer, serine in a protein sequence has been counted as glycine and aspartate as alanine.

Proteinoids produced by heating an equimolar mixture of amino acids at 185°C for 1 hour or by cocondensation of equimolar concentrations of 18 amino acyl adenylates had compositions surprisingly like that of an average protein (Fox and Waehneldt, 1968; Krampitz and Fox, 1969; see Table 4).

We can rationalize the fact that widely different organisms have closely similar amino acid compositions (Fig. 3) (Vegotsky and Fox, 1962) on the basis that their catalytic apparatuses are basically the same. It is not immediately obvious why an abiotic condensation of the amino acids should form a mixture of polypeptides of approximately the same composition as the "average" protein.

PREBIOLOGICAL MODELS: ENZYMES

Not only is it rather easy to produce primitive proteins or proteinoids but these abiotic polypeptides have some catalytic activity. A thermally produced copolymer of 18 amino acids hydrolyzed *p*-nitrophenyl acetate 15 times faster than free histidine or the hydrolyzate of the proteinoid (Rohlfing and Fox, 1967). The histidine residues of the proteinoids were the major contributors to the catalytic activity of the proteinoid. The simultaneous presence in the proteinoid of histidine and of imide linkages seemed to be necessary for catalytic activity. Thermal copolymers of glutamate, aspartate,

TABLE 5

INHIBITION AND REACTIVATION OF HYDROLYSIS OF *p*-NITROPHENYLACETATE[a, b]

Treatment of polymer	Hydrolysis rate[c]	Original rate
No treatment	11.8	100
6 Hours DFP[d]	2.2	18.5
6 Hours DFP followed by 1 hour buffer[e]	2.7	22.8
6 Hours DFP followed by 1 hour TMB-4[f]	9.6	81.5

[a] From Usdin *et al.* (1967).

[b] Reaction conditions: 20 mg polymer suspension per milliliter; *p*-nitrophenylacetate conc. 2×10^{-3} *M*; pH 6.2; 0.007 *M* phosphate; 25°C.

[c] Micromoles of *p*-nitrophenol produced per minute per milliliter of reaction mix.

[d] 1×10^{-3} *M* DFP, pH 6.2; 0.067 *M* phosphate; 25°C.

[e] 0.067 *M* phosphate, pH 6.2.

[f] 1×10^{-1} *M* TMB-4 at pH 6.2; 0.067 *M* phosphate; 25°C.

TABLE 6
ACTIVITY OF MELANOCYTE-STIMULATING HORMONE (MSH) IN THERMAL POLYANHYDRO-α-AMINO ACIDS[a]

Polymer	Activity (units per gram)
Polyanhydro (Ala,Asp,Glu,Leu,Lys,Phe)	0.0
Polyanhydro (Arg,Glu,Gly,Hsd,Phe,Trp)	2.2×10^4
Corresponding free amino acids	0.0
Polyanhydro (Arg,Glu,Gly,Hsd,Phe)	1.4×10^3
Polyanhydro (Glu,Gly,Hsd,Phe,Trp)	0.0
L-Glutamyl-L-histidyl-L-phenylalanyl-L-arginyl-L-tryptophanylglycine	2.2×10^5
α-MSH	3.3×10^{10}

[a] From Fox and Wang (1968).

lysine, histidine, serine, and tyrosine showed Michaelis-Menten kinetics and the activity was inhibited by diisopropylfluorophosphate (Usdin *et al.*, 1967; Table 5). Proteinoids also catalyze the hydrolysis of *p*-nitrophenylphosphate (Oshima, 1968), the decarboxylation of oxaloacetic acid (Rohlfing, 1967), and the decarboxylation of pyruvate (Krampitz and Hardebeck, 1966). The pentapeptide L-threonyl-L-alanyl-L-seryl-L-histidyl-L-aspartic acid put together by a stepwise residue-by-residue procedure has esterase activity (Sheehan *et al.*, 1966).

More astonishing is the report that thermal polymers of arginine, glutamic acid, glycine, histidine, phenylalanine, and tryptophan have melanocyte-stimulating activity (Fox and Wang, 1968; see Table 6).

These model enzyme systems are of interest because they allow serious consideration and experimental evaluation of the notion that primitive enzymes were produced in aqueous solutions in a manner similar to the condensation reactions already mentioned.

What is the possibility that the polymerization of amino acids to form the active site of an enzyme progenitor might be enhanced by the presence of the substrate? A common sequence of a number of esterases and peptidases is Gly-Asp-Ser-Gly. In a nonrestricted coupling of unprotected serine and aspartate one would expect to obtain six different dipeptides only one of which is biologically meaningful. The ratio of the formation of aspartylserine in the presence and in the absence of *N*,*N*-dimethylformamide was 1.4. The same ratio for glycylglycine was 1.15. On the basis of these experimental results it is proposed that "biologically pertinent peptide sequences were produced prebiotically in aqueous solution without the involvement of nucleic acids or residue-specific condensing agents" (Steinman and Cole, 1967).

These results considered along with the report (Steinman and Cole, 1967) that the presence of a polypeptide increases the rate of condensation of amino acid monomers with each other show us that these abiotic systems mimic to a surprising extent reactions characteristic of living cells. It is easy to imagine that polypeptides formed under a given set of conditions in the presence of given "substrates" will reflect those conditions in their amino acid sequences. The sequence constitutes a record of those conditions. But it is not a blueprint for the formation of more identical polypeptides. Identical polypeptides are produced only as long as the conditions remain constant. If a primitive self-replicating polymer molecule found a particular sequence of particular value it could ensure the continued production of this sequence by (1) maintaining conditions constant (that is by providing a barrier that would insulate the condensing system from the changes in the environment), or (2) developing a template system that would specify and select the correct mononers from a changing environment.

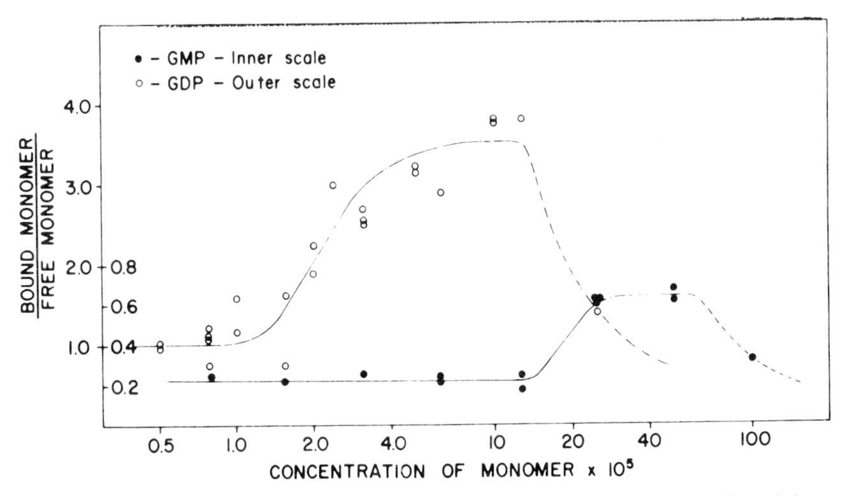

FIG. 4. Equilibrium dialysis of guanosine mono- and diphosphate *vs.* polyarginine. Polyarginine concentration is 1×10^{-3} monomolar. Medium is 0.1 *M* Tris buffer, pH 7.5. Abscissa: concentration of nucleoside mono- or diphosphate; ordinate: ratio of bound monomer to free monomer. From Woese (1968).

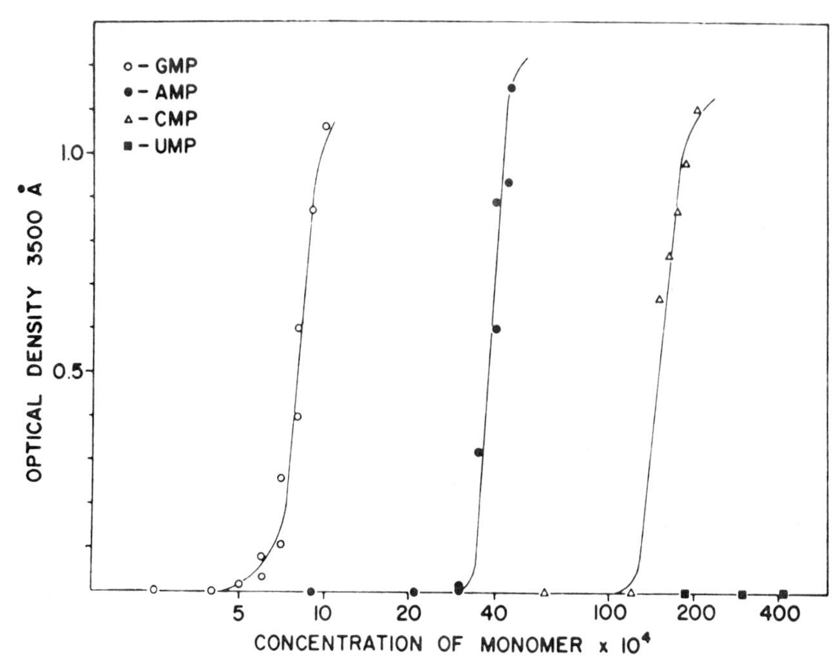

FIG. 5. Formation of insoluble complexes between polyarginine and nucleoside monophosphates. Abscissa: concentration of nucleoside monophosphate; ordinate: turbidity as measured by optical density at 3500 Å, 1-cm light path. Polyarginine concentration is 5×10^{-4} monomolar. Medium is 0.1 *M* Tris buffer, pH 7.5. From Woese (1968).

PREBIOLOGICAL MODELS: REPLICATION

A cell cannot, in spite of its elegant apparatus, reproduce itself if it is deprived of its DNA. Nor can DNA by itself, even if suspended in the midst of all the necessary substrates create the cell for which it contains the blueprints. To expect DNA to accomplish such a thing is roughly comparable to hoping to create Shakespeare by incubating a copy of Hamlet in a warm, moist place. The blueprint is worthless without a shop to read it, and the shop cannot build without a blueprint. For prebiological times to have developed into biological times the elementary blueprint and the simple shop must have evolved simultaneously.

Direct interaction, that is, communication between polypeptides and nucleotide monomers, is possible on a limited scale (Woese, 1968) as shown in Figs. 4 and 5. And a complex between arginine methyl ester and polyguanylic acid will form at concentrations above 10^{-3} M (Woese, 1968). However, such complexes are limited to the basic amino acids. These two systems—basic polypeptides with nucleotide monomers and purine-rich polynucleotides with basic amino acid monomers—together constitute a two-component biopolymer autocatalysis cycle. This colinear 1:1 relationship could serve an accumulation function. It is not a code in the present meaning of the word.

Efforts to demonstrate a stereochemical relationship between the protein amino acids and their codons (Welton and Pelc, 1966) or anticodons (Dunnill, 1966) have not been accepted (Crick, 1967a, b). The origin of the process of translation as we now know it still seems obscure.

I shall show a single example of a kind of communication between a primitive enzyme and its substrates (Knowles and Parsons, 1969; see Table 7). In this model *N*-decylimidazole is the model enzyme. It has a high specificity for the decanoate ester as compared to the acetate ester. This results from the hydrophobic forces that bring the imidazole group close to the ester bond to be hydrolyzed.

EPILOGUE

Let us skip along a few years to the time when real life had appeared. It is agreed that the first life was anaerobic. In the absence of oxygen a relatively high content of sulfhydryl groups in proteins

TABLE 7

RATE CONSTANTS FOR CATALYZED HYDROLYSIS OF *p*-NITROPHENYL ESTERS[a, b]

Catalyst	Acetate ester[c]	Decanoate ester[d]	$k$decanoate/$k$acetate
Hydroxide ion	695	28.6	0.041
N-Ethylimidazole[e]	23	0.83	0.036
N-*n*-Decylimidazole[f]	33	614	18.6
$k$decylimidazole/$k$ethylimidazole	1.43	740	515

[a] From Knowles and Parsons (1969).

[b] Rate constants (M^{-1} min^{-1}) were obtained spectrophotometrically by measuring the rate of appearance of *p*-nitrophenate ion at 400 mμ. All rates were measured at 25°C, in 0.02 M carbonate buffers of pH 9.5 to 10.5, 1% (v/v) acetone.

[c] 3.4×10^{-5} M.

[d] 1.13×10^{-5} M.

[e] 6.2×10^{-4} to 2.9×10^{-3} M.

[f] 1.4×10^{-5} to 7.17×10^{-5} M.

might have been acceptable. Serine and cysteine residues and threonine and thiothreonine residues might well have been used interchangeably in many places in the peptide structure. With increased oxygen tension these extra thiol groups would presumably be less acceptable because of the opportunity for excessive disulfide cross-linking. The cells of pre-aerobic times might also have lacked tyrosine, hydroxyproline and hydroxylysine. The supposition seems especially good for hydroxyproline and hydroxylysine because there are no code words for these two amino acids.

REFERENCES

CLOUD, P. E., JR. (1968). Atmospheric and hydrospheric evolution on the primitive earth. *Science* **160,** 729–735.

CRICK, F. H. C. (1967a). Origin of the genetic code. *Nature* **213,** 119.

CRICK, F. H. C. (1967b). An error in model building. *Nature* **213,** 798.

DUNNILL, P. (1966). Triplet nucleotide-amino acid pairing; a stereochemical basis for the division between protein and non-protein amino acids. *Nature* **210,** 1267–1268.

FERRIS, J. P. (1968). Cyanovinyl phosphate: a prebiological phosphorylating agent? *Science* **161,** 53–54.

FERRIS, J. P., SANCHEZ, R. A., and ORGEL, L. E. (1968). Studies in prebiotic synthesis. III. Synthesis of pyrimidines from cyanoacetylene and cyanate. *J. Mol. Biol.* **33,** 693–704.

FOX, S. W. (Ed.) (1965). "The Origins of Prebiological Systems." Academic Press, New York.

FOX, S. W., and WAEHNELDT, T. V. (1968). The thermal synthesis of neutral and basic proteinoids. *Biochim. Biophys. Acta* **160,** 246–249.

FOX, S. W., and WANG, C. T. (1968). Melanocyte-stimulating hormone: activity in thermal polymers of alpha-amino acids. *Science* **160,** 547–548.

GARRISON, W. M., MORRISON, D. C., HAMILTON, J. G., BENSON, A. A., and CALVIN, M. (1951). Reduction of CO_2 in aqueous solutions by ionizing radiation. *Science* **114,** 416.

HARADA, K., and FOX, S. W. (1964). Thermal synthesis of natural amino acids from a postulated primitive terrestrial atmosphere. *Nature* **201,** 335–336.

HODGSON, G. W., and BAKER, B. L. (1967). Porphyrin abiogenesis from pyrrole and formaldehyde under simulated geochemical conditions. *Science* **216,** 29–32.

HODGSON, G. W., and PONNAMPERUMA, C. (1968). Prebiotic porphyrin genesis: porphyrins from electric discharge in methane, ammonia, and water vapor. *Proc. Natl. Acad. Sci. U. S.* **59,** 22–28.

KNOWLES, J. P., and PARSONS, C. A. (1969). Proximity effect in catalyzed systems: a dramatic effect on ester hydrolysis. *Nature* **221,** 53–54.

KRAMPITZ, G., and FOX, S. W. (1969). The condensation of the adenylates of the amino acids common to protein. *Proc. Natl. Acad. Sci. U. S.* **62,** 399–406.

KRAMPITZ, G., and HARDEBECK, H. (1966). Der durch thermische Protenoide beschleunigte Pyruvat-Abbau in wässriger Lösung. *Naturwissenschaften* **53,** 64–65.

LOHRMAN, R., and ORGEL, L. E. (1968). Prebiotic synthesis: phosphorylation in aqueous solution. *Science* **161,** 64–66.

LOEB, W. (1913). Über das Verhalten des Formamids unter der Wirkung der stillen Entladung. Ein Beitrag zur Frage der Stickstoffassimilation. *Chem. Ber.* **46,** 684–697.

MATTHEWS, C. N., and MOSER, R. E. (1967). Peptide synthesis from hydrogen cyanide and water. *Nature* **215,** 1230–1234.

MILLER, S. L. (1953). A production of amino acids under possible primitive earth conditions. *Science* **117,** 528–529.

MOROWITZ, H. J. (1968). "Energy Flow in Biology." Academic Press, New York.

OPARIN, A. I. (1968). "Genesis and Evolutionary Development of Life." Academic Press, New York.

ORO, J., and GUIDRY, C. L. (1961). Direct synthesis of polypeptides. I. Polycondensation of glycine in aqueous ammonia. *Arch. Biochem. Biophys.* **93,** 166–171.

OSHIMA, T. (1968). The catalytic hydrolysis of phosphate ester bonds by thermal polymers of amino acids. *Arch. Biochem. Biophys.* **126,** 478–485.

PONNAMPERUMA, C., and KIRK, P. (1964). Synthesis of deoxyadenosine under simulated primitive earth conditions. *Nature* **203,** 400–401.

PONNAMPERUMA, C., and MACK, R. (1965). Nucleotide synthesis under possible primitive earth conditions. *Science* **148,** 1221–1223.

PONNAMPERUMA, C., and PETERSON, E. (1965). Peptide synthesis from amino acids in aqueous solution. *Science* **147,** 1572–1574.

RIDER, K., and MOROWITZ, H. J., JR. (1968). The most probable covalent bond distribution in non-equilibrium systems of an atomic composition characteristic of the biosphere. *J. Theoret. Biol.* **21,** 278–291.

ROHLFING, D. L. (1967). The catalytic decarboxylation of oxaloacetic acid by thermally prepared poly-α-amino acids. *Arch. Biochem. Biophys.* **118,** 468–474.

ROHLFING, D. L., and FOX, S. W. (1967). The catalytic activity of thermal polyanhydro-α-amino acids for the hydrolysis of *p*-nitrophenyl acetate. *Arch. Biochem. Biophys.* **118,** 122–126.

SHEEHAN, J. C., BENNETT, G. B., and SCHNEIDER, J. A. (1966). Synthetic peptide models of enzyme active sites. III. Stereoselective esterase models. *J. Am. Chem. Soc.* **88,** 3455–3456.

SHKLOVSKII, I., and SAGAN, C. (1968). "Intelligent Life in the Universe." Dell, New York.

STEINMAN, G., and COLE, M. N. (1967). Synthesis of biologically pertinent peptides under possible primordial conditions. *Proc. Natl. Acad. Sci. U. S.* **58,** 735–742.

STEINMAN, G., LEMMON, R. M., and CALVIN, M. (1965). Dicyandiamide: possible role in peptide synthesis during chemical evolution. *Science* **147,** 1574–1575.

STEINMAN, G., SMITH, A. E., and SILVER, J. J. (1968). Synthesis of a sulfur-containing amino acid under simulated prebiotic conditions. *Science* **159,** 1108–1109.

TYNDALL, J. (1871). "Fragments of Science for Unscientific People." Longmans, Green, London.

USDIN, V. R., MITZ, M. A., and KILLOS, P. J. (1967). Inhibition and reactivation of the catalytic activity of a thermal α-amino acid copolymer. *Arch. Biochem. Biophys.* **122,** 258–261.

VEGOTSKY, A., and FOX, S. W. (1962). Comparisons among heterologous proteins. *In* "Comparative Biochemistry" (M. Florkin and H. L. Mason, eds.), Vol. IV, pp. 185–244. Academic Press, New York.

WEIMANN, B. J., LOHRMAN, R., ORGEL, L. E., SCHNEIDER-BERNLOEHER, H., and SULSTON, J. E. (1968). Template-directed synthesis with adenosine-5′-phospor-imidazolide. *Science* **161,** 387.

WELTON, M. G. E., and PELC, S. R. (1966). Specificity of the stereochemical relationship between ribonucleic acid-triplets and amino-acids. *Nature* **209,** 870–872.

WOESE, C. R. (1968). The fundamental nature of the genetic code: prebiotic interactions between polynucleotides and polyamino acids or their derivatives. *Proc. Natl. Acad. Sci. U. S.* **59,** 110–117.

DEVELOPMENTAL BIOLOGY SUPPLEMENT **3,** 33–58 (1969)

An Analysis of Bacterial Growth

OLE MAALØE

University Institute of Microbiology, Copenhagen, Denmark

1. THE SYSTEM

A growing bacterium is a self-contained system whose dominating activity is protein synthesis. Between four-fifths and nine-tenths of the carbon assimilated, and a similarly large fraction of the energy consumed serve this need. As the mediators of protein synthesis the ribosomes play a key role, and it was therefore not surprising to find that mechanisms exist which permit bacteria to adjust the number of ribosomes they produce in accordance with the environment they grow in. But it was both unexpected and gratifying to see that, by and large, this adjustment follows a simple, almost "sensible" rule, namely, in a given environment, no more ribosomes are produced than can be engaged with high efficiency in protein synthesis. This observation allows some of the major syntheses in these cells to be described in relatively simple terms, and the model to be presented is an attempt to account for a mass of data which are suggestive enough, I think, to warrant this exercise.

2. THE COLLECTION OF DATA

For years our chief interest has been to study the control mechanisms which determine and stabilize the growth rate of bacteria characteristic of a particular medium. To analyze the "fine adjustments" necessary to maintain a steady state of growth, data had to be obtained without perturbing the system by manipulating the cultures. With this restriction, the only variable left to work with was the growth rate itself, and, fortunately, this rate can be varied over a wide range by choosing among different media. Our present model is therefore based critically on measurements of the relative quantities of protein, RNA, and DNA in samples drawn from steady state cultures. The degree to which this ideal state was approximated in our experiments is discussed in a recent monograph (Maaløe and Kjeldgaard, 1966).

Supplementary, but very valuable, information has come from experiments involving shifts between media. By far the simplest experiment of this type is a shift-up which does not impose new synthetic activities on the cells. The best example of such a "gratuitous" shift is one in which amino acids and nucleosides are added to a steady-state culture in a minimal medium with glucose as the only carbon and energy source. The cells immediately increase their growth rate, probably as a direct result of multiple repressions and feedback inhibitions and a corresponding, drastic reduction of a considerable number of synthetic activities. The reverse shift is difficult to interpret. After the shift, the capacity of the cells for synthesizing their own amino acids, etc., is extremely low, and before growth at the definitive postshift rate can be established, the enzyme equipment of the cells has to be readjusted. So far, this slow and gradual process is not well understood.

For later reference the main results of the various measurements are numbered and listed here. If nothing else is said, all quantities and numbers of molecules are normalized to DNA; the reference unit chosen is one genome equivalent of DNA (written: per genome). Standard abbreviations are used throughout, except that the "r" which stands for "ribosomal" in rRNA is also used in the combination r-protein, and as subscript; e.g., in α_r, which designates the r-protein as fraction of all protein. The growth rate, μ, is expressed in doublings per hour.

2.1 The total protein, per genome, has been measured in *Salmonella typhimurium* (see Maaløe and Kjeldgaard, 1966) and in strain TAU-bar of *Escherichia coli* (Forchhammer and Lindahl, in preparation). In both organisms the quantity of protein, per genome, represents 4 to 5 $\times$ 10^8 amino acids. Within experimental error, *this figure is independent of* μ (between $\mu \approx 0.2$ and $\mu = 2.5$).

2.2 The number of ribosomes, per genome, has been calculated from measurements of total RNA after subtraction of tRNA. The original figures for *S. typhimurium* (Maaløe and Kjeldgaard, 1966) have been corrected for errors in the estimates of tRNA (Kjeldgaard, 1967), and the agreement between measurement on different organisms and in different laboratories is now reasonably good. At relatively high growth rates, the number of ribosomes, per genome, is proportional to μ; but for values of $\mu_{37°}$ much below unity the number of ribosomes, per genome, is somewhat higher than would be expected on the basis of strict proportionality. Details about these important relationships are given by Rosset *et al.* (1966), Kjeldgaard (1967), and Forchhammer and Lindahl (in preparation); extensive measurements by R. Lavallé (personal communication) agree with the published data. To indicate the actual numbers obtained it may suffice to state that strain TAU-bar contains close to 10^4 ribosomes, per genome, at $\mu_{37°} = 1.5$.

It should be noted that proportionality between μ and the *number* of ribosomes imply that the *rate* of ribosome synthesis is proportional to μ^2. To illustrate this, compare two steady states with the growth rates μ and $\mu/2$, respectively. The corresponding numbers of ribosomes then are n_r and $n_r/2$, and the rates of synthesis are consequently $n_r\mu$ and $(n_r/2) \times (\mu/2)$. Thus, to a factor 2 between the μ values corresponds a factor 4 between the rates of ribosome synthesis.

Measurements of α_r (Schleif, 1967a, b, 1968) strongly support the notion of proportionality between ribosome number and μ (see Fig. 1).

2.3 As mentioned above, tRNA has been measured as a fraction of the total RNA in several laboratories. The total quantity of tRNA varies little with μ, and a representative number of tRNA molecules, per genome, is 2×10^5.

2.4 The mRNA activity, per milligram of extracted RNA can be measured *in vitro*, and this technique has been used extensively by J. Forchhammer and collaborators in our laboratory. The validity of this assay has been discussed by Forchhammer and Kjeldgaard (1967).

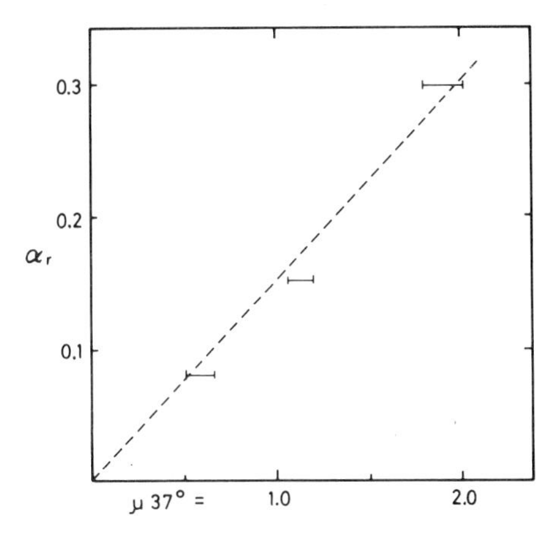

FIG. 1. The α_r and μ values were measured several times in minimal media with succinate or glucose as sole carbon sources, and in a glucose minimal medium enriched with casamino acids, etc. (Schleif, 1967a). The bars illustrate the ranges of growth rates observed.

It is observed that the *in vitro* mRNA activity, per ribosome equivalent of rRNA in the extracts, is nearly independent of μ. This is interpreted to mean that the quantity of mRNA, per ribosome, is nearly the same at all growth rates (Forchhammer and Kjeldgaard, 1968). Like rRNA, mRNA is thus roughly proportional to μ. Note, however, that to maintain this relation the unstable mRNA has to be synthesized at a *rate* which is also proportional to μ (*not* to μ^2, as in the case of the stable rRNA).

2.5. The distribution of the ribosomal material between subunits, single 70 S particles, and polysomes as function of μ has been determined in strain TAU-bar by Forchhammer and Lindahl (in preparation). In this strain, the 30 S and 50 S subunits together make up about 15% of the total at all growth rates. This fraction contains the mature, recycling subunits (Kaempfer *et al.*, 1968), but is probably contaminated with an unknown amount of subunit precursors. Most of the 70 S particles are in the polysome fraction. The *in vivo* distribution of these particles is uncertain since some breakdown of polysome material (and perhaps some runoff of 70 S units) occurs during extraction and preparation. The present estimate is that 70–85% of the total material is in polysomes *in vivo*.

3. NUMERICAL ANALYSIS

Consider first a bacterial culture in a steady state of growth with the growth rate, μ, defined by $M_t = M_0 \exp(c_1\mu t)$ or, $dM/M = c_1\mu dt$, where M is the mass per unit volume of any cell component, and t is time. Since we express μ in doublings per hour, and t in minutes, the constant $c_1 = \ln 2/60$. Let the system contain N amino acids in all of its protein, and $\alpha_r N$ in the r-proteins. The latter reside largely in mature ribosomes with approximately 10^4 amino acids per 7 S particle, and the number of ribosomes in the system is therefore $\approx \alpha_r N \times 10^{-4}$. As a unit of protein synthesis we choose a "*ribosome-minute*," r', which is the average number of amino acids added per minute to a growing polypeptide chain; i.e., the average chain growth rate. This definition gives, $10^{-4} r' = (dN/dt)/(\alpha_r N)$ which, combined with the differential growth equation, yields

$$\mu = c_2 \alpha_r r' \tag{3-1}$$

where the constant c_2 is approximately 8.7×10^{-3}.

As derived here on the basis of protein synthesis in a steady state of growth, equation (3-1) can be applied directly to Schleif's α_r

measurements (cf. 2.2). Figure 1 is a plot of α_r against corresponding values of μ, and the close proportionality between these parameters shows that r' is more or less constant over a considerable range of growth rates. This graphical analysis is rather insensitive to deviations from proportionality at low μ values. It should therefore be restated here that r' decreases significantly at low growth rates (cf. 2.2). This property of the system will be discussed in Section 5.2.

The chain growth rate of a specific protein, the β-galactosidase, has been measured by Lacroute and Stent (1968); they estimate $r'_{37°}$ at about 800–900. Equation (3-1) can therefore be reduced to

$$\mu \approx c_3\alpha_r, \tag{3-2}$$

where $c_3 = c_2r'$ is approximately 8. A system depending on r-proteins only for its growth ($\alpha_r = 1$) would thus double about eight times per hour. A similar figure was quoted years ago by Leslie Orgel on the basis of data from Schaechter *et al.* (1958). As everybody knows, three doublings per hour is about the highest growth rate attainable at 37°C, and it seems that, to achieve this, as much as 35–40% of the cells protein must be r-protein.

Our general knowledge of the mechanism of protein synthesis indicates that a system which obeys equation (3-2) must have the following properties: (a) the production of rRNA must at least match that of r-protein; (b) a constant fraction of all ribosomes must be engaged in protein synthesis; and (c) the various components involved in this process, including the transcribing and activating enzymes, must supply enough mRNA and maintain adequate concentrations of amino acid charged tRNA's. In fact, these inferences are supported by independent, experimental evidence. Thus, Schleif (1968) has demonstrated that growing *E. coli* cells contain very little free r-protein; and we find that the polysomes are made up of a nearly constant fraction of all ribosomes, irrespective of growth rate (cf. 2.5). As regards (c) two points should be made: first, that the mRNA content of growing cells, *per ribosome*, is constant; i.e., the same piece, or length, of mRNA is available to a ribosome whether growth is fast or slow (cf. 2.4). Second, a shift-up from a glucose minimal medium to broth does not measurably increase r' (Maaløe and Kjeldgaard, 1966).

The significance of the last point should perhaps be explained. When broth is added to a minimal medium culture, all the amino acid pools swell, as evidenced by the fact that repression and feedback inhibition become effective at once and as shown directly for a few amino acids by Britten and McClure (1962). When a new amino acid is added to a growing polypeptide chain, it seems obvious that the trial and error process through which the right tRNA is fitted to its codon occupies most of the time involved in the overall reaction. The fact that r' remains unchanged after the shift indicates that the protein synthesizing machinery was fully primed with charged tRNA already *before* the shift; this in turn suggests that the r' value observed is about as high as the physical properties of the components of the system allow.

The relations between r', α_r, and μ were developed formally for an "average" r'. For theoretical reasons the same r' must be assumed to apply to *all* ribosome-mediated protein syntheses, and experiments indicate that r' does not vary during the bacterial division cycle. Thus the same induction kinetics apply for β-galactosidase when inducer is added at different times during the cycle (Cummings, 1965), and autoradiographic studies by Ecker and Kokaisl (1969) indicate that r' is constant in time.

The notion that r' is relatively independent of μ is not new. It was first suggested by Schaechter *et al.* (1958) on the basis of rather crude measurements of total RNA in cells growing at different rates. Since then this type of experiment has been considerably refined, and the data now available permit us to estimate relative as well as absolute values of r'.

First, let us examine the rather striking fact (a) that the total quantity of protein, per genome, is nearly the same at all growth rates. In one doubling time (t min) some 4 to 5 $\times$ 10^8 amino acids must therefore be built into protein, per genome, in the growing culture (cf. 2.1). We also note (b) that the quantity of rRNA and also the number of ribosomal particles, per genome, are more or less proportional to μ (cf. 2.2). Since $\mu = 60/t$, it follows from (b) that the number of ribosome-minutes, *per doubling time*, is nearly constant; taking (a) and (b) together it can be seen that this is true also of r'. It is an interesting property of this system that it always produces about the same quantity of protein, per genome, irrespective of the composition of this protein in terms of enzymes, r-protein, etc. This point will be discussed in some detail in the next sections.

Assuming that all ribosomes are active in protein synthesis, a minimum value can be assigned to r'. Estimates of the number of ribosomes necessary for the actual calculations have been obtained from measurements of rRNA, r-protein, and whole particles, respectively (cf. 2.2 and 2.5).

The minimum values obtained for *S. typhimurium* and given by Maaløe and Kjeldgaard (1966), have been slightly adjusted in the light of the improved tRNA measurements referred to in 2.2. The present conclusion is that the chain growth rate is constant, at about 16 amino acids sec^{-1} at 37°C, at medium and high growth rates (μ = 1.2 and μ = 2.4, respectively), and that it is reduced by about 40% at $\mu \approx 0.2$; thus r' drops from 900–1000 to about 600. With strain TAU-bar, Forchhammer and Lindahl (in preparation) obtained similar figures, and much the same decrease at low μ-values. The same general picture is reported by Rosset *et al.* (1966) except that the constancy at high growth rates is less pronounced, and in some strains r' decreased more or less continuously.

Lacroute and Stent (1968) estimated r' for β-galactosidase. Their figure of 800–900 is also a minimum estimate, since it is based on measuring the time it takes to synthesize the polypeptide chain of the enzyme plus the unknown time required to fold the chain and produce the active tetramere. An r' of about 800 was calculated for the r-proteins by Schleif (1967a, b).

All the minimum estimates based on participation in protein synthesis of all the ribosomes should be corrected so as to express r' in terms of actively engaged ribosomes. According to 2.5 the minimum estimates should be increased by approximately 25%.

4. GENETIC ANALYSIS

We have seen that a steady state of growth is characterized by an α_r value which uniquely defines the growth rate μ as long as r' remains unchanged. Schleif originally used an α to indicate the fraction of all protein which is ribosomal. In the present description of the system, I have called this fraction α_r, to distinguish it from a general α_i, where the subscript refers to any nonribosomal species of protein in the cell. By definition the sum of all α values, including α_r, is unity. A numerical equivalent to α_i is the number of amino acids, per genome, residing in protein "i"; thus an α_i = 0.01 means that the corresponding protein species contain a total of 4 to 5 $\times 10^6$ amino acids (1% of the total).

Our problem is then to understand how the α_r which defines a given steady state is reached, and maintained. The central idea of the model developed here is that α_r is determined by a multivariable function, namely *the entire set of repressions prevailing in the growing cell* (operon-specific as well as catabolite repressions; induction, of course, is viewed as a decrease in degree of repression). A short and incomplete version of the model was published in a note two years ago (Maaløe, 1968).

The concept of a set of repressions requires elaboration. The cistrons representing the different *E. coli* proteins map singly or in groups, corresponding to functionally related enzymes, and transcription is controlled by highly specific effectors (Jacob and Monod, 1961) or by less selective, catabolite effectors (reviewed by Anderson and Wood, 1969). If we focus on a particular unit of control, an operon O_i, we can therefore ask about the probability, P_i, that the next act of transcription in the cell takes place in this rather than any other segment of the genome. If we assume, as is often done tacitly, that all messenger cistrons yield the same average number of protein molecules, then, for a cistron of average length, $P_i = \alpha_i$; i.e., the relative frequency of transcription equals the relative abundance of the final product. Of course, polypeptide chains of very different molecular weight are produced, but the multivariable function we consider represents a large number of separately controlled cistrons. As a good approximation we can therefore think in terms of an "average" cistron producing an "average" quantity of protein.

Focusing again on the operon O_i, we note that three parameters (at least) are involved in determining the yield of the corresponding protein(s). First, the activity of the *operator* which we assume depends on the concentration in the cell of the operator-specific effector, and of the less specific, catabolite effectors; second, the affinity of the *promotor* site for the transcribing polymerase; and, third, the average *gene-dose*. As described here, the DNA structure in the promotor region of the operon determines the efficiency with which a colliding polymerase molecule attaches. This efficiency is thus a permanent, individual property of an operon. The gene-dose is relevant to the model mainly because the multifork pattern of replication, characteristic of rapidly growing cells, accentuates the difference in gene-dose between early and late replicating cistrons.

Of the three parameters described, only the operon control is of immediate interest to us now. Consider first the state of an operon in terms of the fraction, ϵ_i, of total time it is open, i.e., *de*repressed. While the operon is in this state, transcription will be initiated with a frequency which is limited by the concentration of various cytoplasmic elements (cf. Section 5.3), and modulated by the structural properties of the promotor. The wide range over which ϵ_i can be varied is typically shown by inducing the synthesis of β-galactosidase in a cryptic (permease-deficient) strain using several concentrations of a specific inducer (Cohen and Monod, 1957). In this way the steady state of enzyme synthesis can be set anywhere between a maximum, at which some 5% of all newly made protein is β-galactosidase, and a background level, yielding maybe 10^3 times less enzyme. At saturating inducer concentrations, the lac operator probably is totally derepressed and therefore transcribed as frequently as the overall system permits. When the inducer is removed and repression takes over, our index of derepression, ϵ_{lac}, may thus drop from a value near unity to about 10^{-3}. In experiments of this kind all other control indices presumably remain practically unchanged, and the growth rate is in fact changed very little by inducing the cells to produce about 5% of their protein in the form of a dispensable enzyme (Novick and Weiner, 1957).

In a steady state of growth, each protein, and indeed any cell component, increases its mass at the growth rate μ of the culture. Applied to individual cells it is, of course, only meaningful to think of ϵ_i, and of gene-dose, as time averages. However, in the population as a whole a well defined ϵ_i, and gene-dose, can be ascribed to each genetically controlled unit. Finally, if the promotor activity is taken into account, an actual transcription index, $\bar{\epsilon}_i$, can be assigned. This index, taken as a fraction of the sum of all indices, is the probability P_i. The multivariable function referred to above can now be identified as the set of $\bar{\epsilon}$ values, and it can be seen that this set constitutes the *partition-function* according to which the individual species of protein are represented in a steady-state culture. This function has two important properties: (1) it is independent of the *intensity* of transcription, i.e., the partitioning is the same whether the probability of initiating an act of transcription at an open site is high or low; and (2), conversely, the elements of the set can be multiplied by a common factor without changing the partitioning, provided $\bar{\epsilon}_i < 1$ applies throughout (see Fig. 2 and legend).

It now remains to discuss actual values of the $\bar{\epsilon}$ index. We have seen that $\bar{\epsilon}_{lac}$ can be varied over a 10^3-fold range and that enormous amounts of β-galactosidase are produced if the partition function remains more or less unchanged, except that $\bar{\epsilon}_{lac}$ is raised to a high

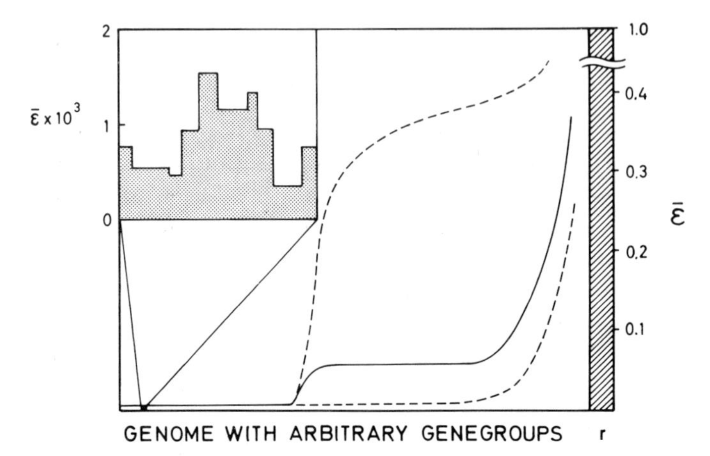

FIG. 2. This graph illustrates the partition-function described in the text as the set of derepression indices $\bar{\varepsilon}_i$.

Ideally each polypeptide encoded in the genome should be represented on the abscissa with a segment corresponding to its molecular weight. From left to right are shown: first the class of enzyme always produced in small quantities (very low $\bar{\varepsilon}$ value); then the class of biosynthetic enzymes with typical $\bar{\varepsilon}$-values around 0.05 in glucose minimal medium (solid curve); followed by a relatively small segment representing proteins assumed to have higher or very high $\bar{\varepsilon}$ values. The r-proteins are represented on the extreme right by the hatched column which, to be clearly visible, has been *increased* five times in width; to be true to scale it should cover about 1% of the genome.

The curves show, qualitatively, the expected shape of the function. The top and center curves represent minimal medium with a "poor" carbon source and with glucose, respectively; the bottom curve shows the maximally reduced partition function expected in a rich medium.

The inset illustrates what a small segment of the curve might look like if the actual molecular weights and $\bar{\varepsilon}$ values were known. As indicated, the segment is chosen from the left-hand part of the graph.

value. Very similar data exist for another inducible enzyme, alkaline phosphatase.

In the light of these examples let us examine some biosynthetic enzymes about which it is known that the activity assays commonly used measure actual quantities of the specific proteins. The three cases I have chosen are: the aspartic transcarbamylase, ACTase, studied particularly by Gerhart and Schachman (see, e.g., Gerhart and Holoubek, 1967); the ornithine transcarbamylase, OTCase, extensively studied by Gorini and his associates (see, e.g., Jacoby and Gorini, 1969); and the tryptophan-synthesizing enzymes, analyzed by Yanofsky's group (see, e.g., Yanofsky and Ito, 1966). Wild-type organisms growing in a minimal medium produce all these enzymes in *small* quantities compared to those produced by mutants in which little or no repression is exerted.

The ACTase normally constitutes 0.1 to 0.2% of the cell protein, but strains exist in which this percentage is more than 20 times higher (J. C. Gerhart, personal communication). In the case of OTCase a similar difference has been demonstrated between wild-type and repression-defective mutants; in the wild type, further, and almost complete suppression of enzyme synthesis is caused by adding arginine to the medium (see Fig. 3 and legend). Finally, a special case recently analyzed by C. Yanofsky (personal communication) should be mentioned. The enzymes of the trp operon constitute about 0.4% of the cell protein in a wild-type, minimal medium culture, but a strain which produces about 15 times more has been isolated. This mutant carries asparagine instead of one of the glycines in the B protein, and this substitution seriously affects the activity of the enzyme. This case is particularly clear, because it shows how the normal, unimpaired control system reacts by derepressing the trp operon, because the efficiency of one of its enzymes is greatly reduced.

The three cases discussed here indicate that enzymes which carry heavy biosynthetic loads, as a rule, are produced under considerable internal repression. In other words, during growth with a single car-

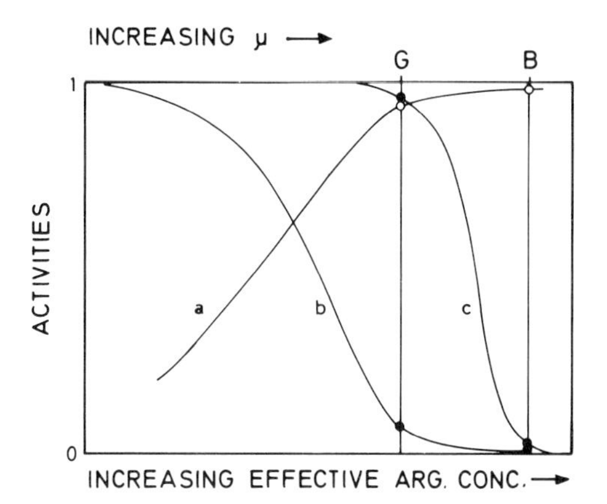

FIG. 3. The three curves illustrate the dependency on the intracellular arginine concentration of (*a*) the charging of $tRNA_{arg}$; (*b*) the degree of derepression of the arginine-synthesizing enzymes; and (*c*) the activity of these enzymes as affected by end-product inhibition. The construction of the curves is described in the text.

bon source the cells seem to maintain pool levels of arginine and tryptophan which cause strong repression. The data indicate that ϵ_{arg} and ϵ_{trp} are reduced to about 1/20, or less, of their maximum values (see Fig. 3).

The low ϵ-values which seem be characteristic of many biosynthetic enzymes are pertinent to this analysis in two ways: on the one hand, here is a group of at least 100 enzymes which make up a large fraction of the total protein in a minimal medium culture, and the synthesis of which can be drastically reduced by amino acids, purines, and pyrimidines added from outside. On the other hand, the same group of enzymes would continue to be synthesized, *in more or less the same proportions*, if all the relevant ϵ values were multiplied by a common factor. This factor could be fairly large without violating the condition that the individual ϵ_i remains below unity. The effect of such a change on the relative abundance of proteins whose ϵ_i did *not* change is discussed in Section 5.2.

Finally, I want to mention a somewhat ill-defined but important class of proteins, namely, those that seem never to be present in the cell in more than a few copies. All the repressor proteins probably belong in this class, and a fair number of enzymes which, in contrast to those discussed above, carry very light biosynthesis loads, may be included (e.g., the enzymes catalyzing the synthesis of the B vitamins). It is convenient to imagine that very "tough" promotors exist, which, without the intervention of other control elements, would keep the frequency of transcription of the cistrons corresponding to this class of proteins very low. In the case of the repressor protein of the lac operon, one mRNA produced per doubling time would seem to be quite adequate (Gilbert and Müller-Hill, 1967). In the terminology used here, the cistron coding for this protein would be characterized by an ϵ value of unity, *and* an $\bar{\epsilon}$ value of, say, 10^{-3} or 10^{-4}.

5. THE MODEL

5.1. *General Properties*

About ten years ago our work on bacterial growth could be summarized in this way:

> The sketch of the growing bacterium presented here is based essentially on the idea of exchange of information between different molecular levels of organization in the cell. A flow of information is assumed to descend from a linear, genetic specification on a DNA strand, *via* RNA and protein, and to give to a small molecule, such as an amino acid, its three-dimensional individuality. Equally specific information is believed to pass from the level of the small molecules back in the

direction of the nucleus. This feedback of information, which produces the phenomenon of repression, is thought to be responsible for one of the remarkable properties of the cell: its ability to adjust the size and activity of the different synthetic systems to the set of nutrients present in the medium; an adjustment which results in the establishment of a definite partitioning of energy and matter among the synthetic systems, to which corresponds a definite growth rate and cell composition. (Maaløe, 1960.)

When this was written, the mechanism of protein synthesis, and the built-in controls, were poorly understood, and many statements were necessarily vague. Today a more detailed analysis can be made. However, our general attitude to the problem has not changed. It still seems necessary to me to treat the bacterial cell as a unit system, and it is attractive to consider the controls of the synthesis of all the individual proteins as equivalent; i.e., as contributing in identical manner, but with different weight, toward determining the growth rate. The present model thus further develops the old idea of describing the unit system in terms of "flow of information."

Much of the data used here are relatively new, and I shall therefore begin the description of the model by restating the basic facts for which it has to account:

(a) At medium and high μ values the number of ribosomes is proportional to μ. This implies that the *rate* of synthesis of the ribosomes is proportional to μ^2. At low growth rates the ribosome numbers are somewhat higher than expected on the basis of strict proportionality (cf. 2.2).

(b) The quantity of mRNA, per ribosome, is nearly constant (cf. 2.4). These unstable RNA molecules do not accumulate as ribosomes do, and if we assume that the messenger half-life is more or less independent of μ, mRNA must be synthesized at a rate proportional to the ribosome number. With the restriction mentioned in (a), the rate of mRNA synthesis is therefore proportional to μ.

(c) The synthesis of rRNA closely matches that of the r-proteins, and the cells maintain very small pools of free rRNA, and of free r-protein (cf. 2.2).

Very briefly, the model accounts for these observations in the following way: the partition-function described in Section 4, and illustrated in Fig. 2, is thought *by itself* to generate α_r, and the balance between ribosomes and mRNA on the one hand, and between r-protein and rRNA on the other, are thought to be maintained by separate feedback mechanisms. These basic elements of the model will now be discussed one by one.

5.2 *Passive, or Indirect Control of* α_r

The full weight of generating α_r can be put on the partition function by having the r-protein segment(s) of the genome always derepressed, i.e., subject to promotor activity, and to the effect of changes in gene dose, but with an $\epsilon_r = 1$. This situation is illustrated in Fig. 2, where the column on the right-hand side represents the ribosomal proteins, and α_r is the area of this column as fraction of the total area under one of the curves (see legend). There can be no doubt that this total area (and thereby α_r) *changes* from one growth condition to another, but it is perhaps not obvious that it changes in the right direction, let alone that it changes sufficiently to account for the observed variations in α_r.

To examine this problem, consider first growth in glucose minimal medium. In *E. coli* and many other organisms, glucose or one of several related compounds, are the carbon and energy sources on which these microorganisms grow fastest. Thus, flow rates of carbon and energy can be established via the glucose pathway(s), which are higher than those obtainable through pathways utilizing other substrates. I emphasize this because it means that growth rates higher than $\mu_{glucose}$ can be obtained only by supplying ready-made building blocks (e.g., amino acids) in the medium, whereas rates below $\mu_{glucose}$ can be realized with carbon sources which are converted less efficiently than glucose (similarly, μ can be reduced by replacing ammonia by other nitrogen sources, but we have made little use of this alternative). The steady state of growth with glucose as the sole carbon and energy source is characterized by an $\alpha_r \approx 0.15$, and a $\mu_{37°} \approx 1.2$; in Fig. 2 the corresponding partitioning between ribosomal and nonribosomal protein is indicated by the solid curve.

In broth, α_r and μ are typically twice as high as in glucose minimal medium. Two factors seem to contribute independently to this increase. In the first place, production of the biosynthetic enzymes is more or less totally repressed, and in view of the reduced energy requirement (per milligram of protein produced) other enzyme systems may be reduced in size. Secondly, the r-protein (at least the "30 S-proteins") seem to map in the early replicating segment of the genome, and the gene-dose would therefore increase from about 1.4 in the glucose culture to about 1.9 in the rich medium (assuming 2 gene copies per 1.4 genome equivalents of DNA in the glucose culture, and 4 copies per 2.1 equivalents in rapidly growing cells; see e.g., Helmstetter *et al.*, 1968). This effect on the gene dose would raise the α_r of a glucose culture from 0.15 to 0.20, and to reach the value of 0.30 observed in the rapidly growing culture, repression would have to reduce the sum of the $\bar{\epsilon}_{glucose}$ values by about 40%. Considering that the biosynthetic enzymes alone probably account for 20–40% of the protein in a glucose culture, this does not sound unreasonable.

The decrease in α_r and μ observed with "poorer" carbon sources may also be accounted for by changes in the partition pattern. To explain a decrease in α_r from 0.15 to, say, 0.05 the sum of the $\bar{\epsilon}_{glucose}$ values must increase threefold. Part of this increase would come from derepressing the synthesis of enzyme systems required by the new carbon and energy source. Since, by definition, this compound is converted less efficiently than glucose, the overall system would be expected to compensate by producing large quantities of the required enzymes, i.e., to derepress the relevant operon(s) more or less completely. However, even complete derepression of a few operons can hardly be imagined to increase the sum of the $\epsilon_{glucose}$ values threefold. It is therefore important that the biosynthetic enzymes represent a large number of operons which can be derepressed to considerable degrees without much change in the partition pattern in that sector (see description of the partition function in Section 4). As we have seen, $\epsilon_{glucose}$ values around 0.05 seem to be typical of enzymes of this class, and the entire threefold increase in the sum of the $\epsilon_{glucose}$ values could probably be achieved by raising all the 5% values to about 50%.

When equations (3-1) and (3-2) were developed, we noted that, for values of $\mu_{37°}$ much below unity, α_r decreases less than predicted by the simplified equation, $\mu = c_3\alpha_r$. Figure 3 is constructed to illustrate how this trend can be understood in terms of the amino acid pool-levels in the cells. The three curves represent: (a) the degree of charging of tRNA, (b) the level of *de*repression, and (c) the level of enzyme activity; all three, of course, are functions of the effective, intracellular concentration of an amino acid (in this case, arginine). The shapes of these curves are largely unknown, but their relative positions can be deduced as follows. In a glucose culture (position *G* on the abscissa), the arginine concentration is sufficient to repress OTCase synthesis to about 5% of its maximum value; at the same position little, if any, end-product inhibition is exerted (one point on curve *b*, and one on curve *c*). The $tRNA_{arg}$ must be more or less fully charged at position *G*, since the pool level can be raised (by adding arginine from outside) without significantly increasing r' (one point on curve a). In a broth culture (B on the abscissa) intense repression and end-product inhibition prevail (low points on curves b and c), and the charging of the tRNA is, if anything, more complete than before (second point on curve a). At concentrations below position G, all three arginine-dependent effects must fade out gradually.

According to the model, α_r decreases as a consequence of general derepression, i.e., as the sum of the ϵ values increases. Thus, low α_r and μ values are supposed to reflect low concentrations of intracellular amino acids, and other quantitatively important effectors. At some point along this line the degree of charging of tRNA must also begin to decrease significantly. The effect of this will be to reduce r', because the average time it takes to add a new amino acid to a growing polypeptide chain will increase. The equation $\mu = c_2\alpha_r r'$ (3-1), which must now be applied, shows that when r' begins to decrease, μ must be expected to decrease more sharply than α_r. This, I believe, is the reason why the proportionality between α_r and μ breaks down at low growth rates (cf. 2.2 and 5.1).

As discussed in Section 4, the amino acid pools definitely increase between positions G and B in Fig. 3; but, we do not yet know to what extent, nor how generally, these pools are reduced at lower growth rates. At the moment more is known about the riboside-triphosphate concentrations which have been shown to decrease more or less linearly with μ (see below). One of them, UTP, acts as corepressor of ATCase synthesis and thus belongs in the group of "other important effectors."

So far our arguments show that the partition function might account for the known variations in α_r without invoking active or direct control of the synthesis of the r-proteins. However, before this main aspect of the model is accepted as a serious hypothesis two critical questions should be examined: (a) Would passive control exerted by the collective of repressions confer stability on the system? (b) Can more or less obvious alternatives be excluded?

The first question can be answered by considering a cell with the ideal composition corresponding to a particular steady state of growth. At division, let the two sister cells receive $(n_r/2) \pm \Delta n_r$ ribosomes, respectively, everything else being evenly distributed. In the cell with excess ribosomes the tendency will be to raise the rate of protein synthesis above the average; this will cause excessive drain on the pools of amino acids, etc., lower their concentrations, and relieve internal repression to some extent. The net result will therefore be to *reduce* α_r temporarily, and thus to bring the ribosome number down. By the same reasoning it can be seen that α_r will increase temporarily in the sister cell. This is the kind of "fine adjustment" referred to in Section 2, where the principles of data collecting were discussed.

The second question cannot be answered definitively, and only one plausible alternative has occurred to me. Suppose the observed changes in α_r could *not* be accounted for without including in the model a control acting directly on $\bar{\epsilon}_r$. To introduce specific repression of the synthesis of the r-proteins (i.e., to allow ϵ_r to assume values below unity) implies the existence of an effector which would contribute to the setting of α_r. We know that the metabolic pattern changes greatly with the carbon source used, and one therefore has to look for an effector whose concentration nevertheless would vary in a monotonic manner with μ.

J. Neuhard (personal communication) has shown that the riboside triphosphate concentrations increase more or less linearly with μ, probably reflecting a parallel increase in the rate at which energy is consumed. One of these triphosphates, or a derivative thereof, could therefore be the effector we look for. However, I think this possibility can be excluded, because experiments show that α_r can be increased and, at the same time, the concentrations of all the riboside triphosphates drop. This situation obtains after a shift from glucose minimal to a rich medium, when α_r actually overshoots its definitive value (Schleif, 1967a), while the triphosphates are temporarily reduced to about a third of their preshift concentrations (unpublished data of J. Neuhard and J. Ingraham, discussed by Maaløe and Kjeldgaard, 1966).

5.3 *The Balance between mRNA and Ribosomes*

The feedback mechanism thought to maintain this balance was first suggested by Stent (1964). The simplest version of Stent's idea is that a ribosome, or one of its subunits, is required to initiate the synthesis of any mRNA molecule. This view is supported by *in vitro* studies of RNA synthesis (Shin and Moldave, 1966); and the work of Revel *et al.* (1968) suggests that the 30 S subunit is the ribosomal element involved in the act of initiation. It is therefore important to recall that the free subunits constitute a fixed fraction (about 15%) of all the ribosomal material (cf. 2.5).

Our present model specifically states that all classes of mRNA, including the r-protein messenger, are synthesized by the same polymerase, *and* that the initiation mechanism is the same. The frequency of initiation is supposed to be proportional to the number of ribosomes at all values of μ; i.e., neither shortage of polymerase, nor queueing along the DNA template is allowed to affect the frequency. In a system with these properties, the initiation frequency will be proportional to the ribosome number, and so, of course, will the number of mRNA molecules produced per minute. Finally, to produce the observed almost constant ratio between the *quantity* of mRNA and the number of ribosomes, the messenger half-life must be assumed to be more or less independent of μ.

Recent measurements of the RNA chain growth rate (Bremer and Yuan, 1968; Manor *et al.*, 1969) show that the number of codons transcribed per second agrees reasonably with the number of amino acids added to a polypeptide chain in the same time. Transcription and translation *could* therefore be intimately coupled. In fact, Manor *et al.* show that, like our r', the RNA chain growth rate is moderately reduced in slow-growing cells.

As an alternative, the frequency of transcription could be imagined to be governed by the polymerase concentration. If the polymerase cistrons were exempt from repression, as we imagine the r-protein cistrons to be, polymerase and ribosomes would always be produced in the same relative amounts. The reason I do not think this scheme would be satisfactory is that a single polymerase seems to be responsible for the synthesis of all three classes of RNA [a point mutation can make the polymerase resistant to rifamycin *in vitro* (R. Schleif, personal communication); and this drug inhibits both mRNA and stable RNA synthesis *in vivo* (B. Watson, personal communication)]. We know that at reasonably high growth rates, d(mRNA)/dt, and d(tRNA)/dt are proportional to μ, whereas d(rRNA)/dt increases as μ^2. This means that the total number of growing RNA chains in a cell, which equals the number of polymerase molecules engaged, increases *more* than linearly with μ. If the polymerase were to limit the frequency of transcription at all growth rates, its concentration would therefore have to increase *more* than does the number of ribosomes. This could not be achieved simply by exempting the polymerase cistrons from repression.

This discussion of mRNA synthesis shows how badly we need good measurements of mRNA half-life, and of polymerase concentration at different, and especially at low growth rates.

5.4 *The Balance between Ribosomal Protein and rRNA*

This part of the model rests quite heavily on Schleif's data (1967a,b, 1968). They are given this weight for two reasons: first, the technique unambiguously separates r-protein from all other proteins in the cell; and, second, Schleif shows that the protein associated with the excess rRNA produced in the presence of chloramphenicol is largely nonribosomal. Together with the kinetic data, this demonstration makes clear that it was wrong to conclude from the existence of the so-called chloramphenicol particles that normally growing cells maintain a large pool of free ribosomal proteins (Kurland and Maaløe, 1962).

To explain the production of matching quantities of r-protein and rRNA at all growth rates, a completely *ad hoc* feature has been introduced into the model: it is assumed that one of the r-proteins acts as an inducer of rRNA synthesis. The hypothetical inducer is thus constantly being introduced into the cytoplasm and removed again by incorporation into new ribosomes.

This scheme obviously serves the purpose for which it was invented, and thereby overcomes a difficulty which may not be too apparent. The extreme diversity among the r-proteins (Moore *et al.*, 1968), plus the evidence that rRNA maps apart from at least some of the r-proteins (experiments by Atwood, communicated by S. Spiegelman), makes it more and more unlikely that nascent rRNA serves as messenger for more than a few, if any, of the r-proteins. An ordinary mRNA molecule yields some 50–100 protein molecules per cistron, and applying this figure to the synthesis of the r-proteins it appears that for each mRNA produced a fairly large number of rRNA molecules are required to balance the protein yield. Even allowing for a four- to sixfold duplication of the rRNA cistrons, it seems that the promotor activity would have to be ten or more times higher in the rRNA than in the r-protein cistrons, and it is conceivable that matching quantities of their products could be assured simply by choosing the proper ratio between the promotor activities (an extra variable with no strings attached can do almost anything for you).

The proposed induction scheme was preferred because it adds a desirable property to the model: *the synthesis of rRNA, but not of mRNA, is put under stringent control.* It has been argued that if a single polymerase carried out all transcription, the three classes of RNA might be affected more or less equally (coordinately) under amino acid starvation (Maaløe and Kjeldgaard, 1966; Edlin and

Maaløe, 1966; Friesen, 1966). There is now strong evidence that mRNA is produced in considerable quantities in amino acid-starved cells (Edlin *et al.*, 1968; Lavallé and De Hauwer, 1968; Morris and Kjeldgaard, 1968; Stubbs and Hall, 1968), and at least qualitatively, this noncoordinate regulation of RNA synthesis agrees with the model. The quantitative aspect is difficult to assess (see Section 6).

6. CONSEQUENCES OF THE ANALYSIS

To construct the skeleton model just described, it was not necessary to decide whether the mechanism thought to balance r-protein and rRNA synthesis represented positive or negative control; nor did we have to know in detail about the coupling between ribosomes and mRNA synthesis. Both mechanisms can probably best be tested *in vitro* by adapting and refining already existing systems for RNA and protein synthesis.

The strength of the model is that it accounts for a large body of quantitative measurements made on *cells in well defined states of growth*. Its weakness is that it is hard to test its main thesis, because the complexity of the system renders the interpretation of even simple *in vivo* experiments ambiguous. Alternative models can almost certainly be constructed; however, it would be difficult to consider such a model seriously unless it included an account of our data, or questioned their validity.

The assembly of repressions, which is described here by the multivariable partition function, must figure in any analysis of bacterial growth. In the present model, this function is assigned special properties with regard to determining α_r, and stabilizing it during steady-state growth. These properties are such that the α_i of a protein generated from a nonrepressible cistron must always, even through a shift experiment, be proportional to α_r (except for possible changes in the relative gene-doses, as discussed in Sections 4 and 5.2). The model would be strongly supported if this prediction were verified. However, the critical test is complicated by catabolite repression. It is a simple matter to select a strain lacking a specific repressor protein, and thus exempt from specific repression of a given operon; but it is difficult to prove that the same operon is insensitive to catabolite repression. Moreover, it may be necessary to insist on *total* insensitivity, since even considerably reduced sensitivity could obliterate relatively small but critical differences. At the moment, the most promising material for this kind of experiment would seem to be strains of the types described by Silverstone *et al.* (1969) which carry mutations assumed to reduce or abolish the sensitivity of the lac-promotor region to catabolite repression. In fact, some of the results reported in their paper point in the direction predicted by our model; others would have to be interpreted in terms of residual, slight sensitivity to catabolite repression.

A less stringent test would be to compare, say, OTCase synthesis in the wild type and in repression-defective mutants at different growth rates. If the mutants chosen were truly nonrepressible, the enzyme level in the wild type would be expected to approach that in the mutants as the growth rate decreased. However, the degree to which the difference would be reduced cannot be predicted (see discussion of growth with "poor" carbon sources in 5.2).

More indirectly the model can be tested by applying it to special cases. Two can be mentioned here:

(a) A relatively small number of different strains, and of different steady states have been carefully examined, and we therefore wanted directly to test one of the consequences of equations (3-1) and (3-2), namely, that for a given μ, the same α_r should obtain irrespective of the carbon and energy source (in minimal media). For this purpose strains were selected carrying a single mutation which greatly reduces the capacity for uptake of a variety of sugars and amino acids. In batch cultures of these strains μ is defined by the concentration of the carbon source, and steady states of growth can be maintained at quite satisfactory cell densities. With lactose as the sole carbon source μ values between 1.5 and 0.5 were established and the RNA:DNA ratios determined by chemical analyses. Throughout, the correspondence between this ratio and μ was almost exactly the same as that found by comparing *different* carbon sources, such as glucose, glycerol, succinate, and acetate. (This study, including the strain selection, was made by K. v. Meyenburg in our laboratory, and will be published elsewhere.)

(b) The second case is clearly pathological, and concerns a mutant analyzed by MacDonald *et al.* (1967). This strain is characterized by an abnormally high RNA:DNA ratio and it grows slowly in all media. The lesion affects the maturation of the 50 S ribosomal subunit, and large pools of precursor material and of free, apparently normal 30 S subunits are maintained during growth. It is obvious that, relative to the low growth rate, these cells contain very large total quantities of r-protein as well as rRNA. In terms of the equation $\mu = c_2\alpha_r r'$ (3-1),

α_r is abnormally high and r' correspondingly reduced. However, equation (3-1) presupposes that all, or a constant fraction of the ribosomal material is in active 70 S particles. Relative to the wild type, this condition is far from being realized in the mutant, and the conclusion about r' being reduced is therefore trivial. The case is brought up to emphasize that if, for any reason, r' is reduced, α_r must be correspondingly increased. This would seem to apply specifically to growth with restricted availability of an amino acid (or of some species of charged tRNA), because the primary effect of the restriction must be to increase the step time for the amino acid in short supply and thus reduce r'. Again, however, the complexity of the system makes it impossible to predict *how much* r', and α_r, would be affected.

7. DEFICIENCIES IN THE ANALYSIS

The phenomenon of "relaxedness" and the control of tRNA synthesis and of DNA replication have deliberately been left out of the main discussion. I shall briefly explain why, and try to relate each of these topics to the model.

It has been shown how stringent control of rRNA synthesis is built into the present model (as it should be). The "relaxed" (RC^-) mutants do not figure in the body of the text because I now believe that they represent a secondary defect in the RNA control mechanism, and tell nothing about its main principle of operation.

Little attention has been given to the fact that amino acid starvation, i.e., the condition in which the RC^- phenotype is revealed, greatly upsets the metabolism of the cells. During normal growth most of the carbon and energy are consumed in protein synthesis, and this flow is cut drastically when a required amino acid is withdrawn. As a result precursors and catabolites must accumulate. Cashel and Gallant (1969) have shown that rare, or abnormal nucleotides accumulate in RC^+ but not in RC^- cells. Unfortunately, neither the work of Gallant and his associates, nor the coupling scheme proposed here, appear to explain why chloramphenicol, and other antibiotics interfering with protein synthesis, uncouple rRNA production in RC^+ cells.

In retrospect, it was probably an error to imagine that studies of RC^- mutants (Stent and Brenner, 1961) and of the uncoupling effect of chloramphenicol (Kurland and Maaløe, 1962) would lead to an understanding of the *main* features of the control of RNA synthesis.

It has long been known that RC^+ and RC^- cells have identical growth characteristics in different media and respond the same way to a shift-up (Neidhardt, 1963). These facts are not readily explained by the old hypothesis which made uncharged tRNA the *main* effector in the control of RNA synthesis, and described RC^- cells as being relatively insensitive to its inhibitory effect on RNA synthesis. The present model allows RC^+ and RC^- cells to respond identically to fluctuations in the partition function, and thus to carry out the fine adjustments required to establish and stabilize a steady state of growth; the two types are thought to differ in a secondary feature of the control system, the coupling between the syntheses of r-protein and rRNA, and this difference is revealed only under conditions of metabolic congestion. In summary, the uncoupling of rRNA (and tRNA) synthesis in amino acid starved RC^- cells, *and* in RC^+ cells in the presence of chloramphenicol, etc., remains somewhat mysterious, despite the impressive amount of biochemical and genetic data now available (reviewed by Edlin and Broda, 1968).

The synthesis of tRNA and replication of DNA have an important characteristic in common: per genome, the same total quantity is produced, during one doubling time, at all growth rates. This is self-evident in the case of DNA, and since the tRNA:DNA ratio is more or less independent of μ it applies to tRNA as well (cf. 2.3).

We have seen that the structure of the whole system imposes the same rule on protein synthesis, i.e., that irrespective of its composition the total quantity of protein per genome is constant. The same could be true within the class of tRNA molecules, but one crucial piece of information is lacking, without which it is impossible to tackle the problem: it is not known whether the individual tRNA species are produced in constant molar ratios, or whether these ratios are subject to specific regulations. Constant molar ratios seem most reasonable to me, because the trial and error process by which the correct species of changed tRNA is selected at any step in polypeptide synthesis probably is governed by diffusion. It would therefore seem that the *concentrations* of the individual species cannot be allowed to change much with the growth rate. However, the problem of the ratios should be settled before speculating further.

Much has been learned in recent years about DNA replication and about its relation to bacterial growth. The pertinent work is well represented in Volume 33 of the Cold Spring Harbor Symposia (1968). In relation to the model, the most important fact is that, like tRNA and total protein, the initiator(s) of replication must be produced in a fixed amount per genome and per doubling time. Very likely, one or more proteins are involved specifically in the initiation of replication, and these proteins must therefore be synthesized as a constant fraction of all the cells proteins, *irrespective of the composition of this assembly*. No simple mechanism for achieving this has suggested itself, and a link is therefore missing between the present model and the elegant scheme constructed by Helmstetter and Cooper to account for the pattern of replication at different growth rates (see Helmstetter *et al.*, 1968).

ACKNOWLEDGMENTS

A large number of colleagues have contributed to the body of data and ideas, which I have attempted to fuse into a model. A complete list of names would be difficult to compose, but I want particularly to thank my old friends, N. O. Kjeldgaard and M. Schaechter, with whom this work was begun some twelve years ago and who have contributed to it ever since.

It should also be emphasized that many ideas appearing in this paper took shape during long discussions with friends both at home and abroad. I wish to thank all of you, and I hope you enjoyed our discussions half as much as I did.

Our work has been generously supported by the National Institutes of Health, U.S.P.H.S. through research grants (E. 3115, AI-04914) to the author.

REFERENCES

ANDERSON, R. L., and WOOD, W. A. (1969). Carbohydrate metabolism in microorganisms. *Ann. Rev. Microbiol.* **23,** in press.

BREMER, H., and YUAN, D. (1968). RNA chain growth-rate in *Escherichia coli*. *J. Mol. Biol.* **38,** 163–180.

BRITTEN, R. J., and MCCLURE, F. T. (1962). The amino acid pool in *E. coli*. *Bacteriol. Rev.* **26,** 292–335.

CASHEL, M., and GALLANT, J. (1969). Two compounds implicated in the function of the RC gene of *Escherichia coli*. *Nature* **221,** 838–841.

COHEN, G. N., and MONOD, J. (1957). Bacterial permeases. *Bacteriol. Rev.* **21,** 169–194.

CUMMINGS, D. J. (1965). Macromolecular synthesis during synchronous growth of *Escherichia coli* B/r. *Biochim. Biophys. Acta* **95,** 341–350.

ECKER, R. E., and KOKAISL, G. (1969). Synthesis of protein, ribonucleic acid, and ribosomes by individual bacterial cells in balanced growth. *J. Bacteriol.* **98,** 1219–1226.

EDLIN, G., and BRODA, P. (1968). Physiology and genetics of the "ribonucleic acid control" locus in *Escherichia coli*. *Bacteriol. Rev.* **32,** 206–226.

EDLIN, G., and MAALØE, O. (1966). Synthesis and breakdown of messenger RNA without protein synthesis. *J. Mol. Biol.* **15,** 428–434.

EDLIN, G., STENT, G. S., BAKER, R. F., and YANOFSKY, C. (1968). Synthesis of a specific messenger RNA during amino acid starvation of *Escherichia coli*. *J. Mol. Biol.* **37,** 257–268.

FORCHHAMMER, J., and KJELDGAARD, N. O. (1967). Decay of messenger RNA *in vivo* in a mutant of *Escherichia coli* 15. *J. Mol. Biol.* **24,** 459–470.

FORCHHAMMER, J., and KJELDGAARD, N. O. (1968). Regulation of messenger RNA synthesis in *Escherichia coli*. *J. Mol. Biol.* **37,** 245–255.

FRIESEN, J. D. (1966). Control of messenger RNA synthesis and decay in *Escherichia coli*. *J. Mol. Biol.* **20,** 559–573.

GERHART, J. C., and HOLOUBEK, H. (1967). The purification of aspartate transcarbamylase of *Escherichia coli* and separation of its protein subunits. *J. Biol. Chem.* **242,** 2886–2892.

GILBERT, W., and MÜLLER-HILL, B. (1967). The lac operator is DNA. *Proc. Natl. Acad. Sci. U.S.* **58,** 2415–2421.

HELMSTETTER, C., COOPER, S., PIERUCCI, O., and REVELAS, E. (1968). On the bacterial life sequence. *Cold Spring Harbor Symp. Quant. Biol.* **33,** 809–822.

JACOB, F., and MONOD, J. (1961). Genetic regulatory mechanisms in the synthesis of proteins. *J. Mol. Biol.* **3,** 318–356.

JACOBY, G. A., and GORINI, L. (1969). A unitary account of the repression mechanism of arginine biosynthesis in *Escherichia coli*. I. The genetic evidence. *J. Mol. Biol.* **39,** 73–87.

KAEMPFER, R. O. R., MESELSON, M., and RASKAS, H. J. (1968). Cyclic dissociation into stable subunits and re-formation of ribosomes during bacterial growth. *J. Mol. Biol.* **31,** 277–289.

KJELDGAARD, N. O. (1967). Regulation of nucleic acid and protein formation in bacteria. *Advan. Microbial Physiol.* **1,** 39–95.

KURLAND, C. G., and MAALØE, O. (1962). Regulation of ribosomal and transfer RNA synthesis. *J. Mol. Biol.* **40,** 193–210.

LACROUTE, F., and STENT, G. S. (1968). Peptide chain growth of β-galactosidase in *Escherichia coli*. *J. Mol. Biol.* **35,** 165–173.

LAVALLÉ, R., and DEHAUWER, G. (1968). Messenger RNA synthesis during amino acid starvation in *Escherichia coli*. *J. Mol. Biol.* **37,** 269–288.

MAALØE, O. (1960). The nucleic acids and the control of bacterial growth. *Microbial Genetics, 10th Symp. Soc. Gen. Microbiol.* (W. Hayes and R. C. Clowes, eds.), pp. 272–293.

MAALØE, O. (1968). Messenger and ribosomal RNA synthesis *in vivo*. *In* "Biochemistry of Ribosomes and Messenger-RNA" (Symposium, Castel Reinhardsbrunn) (R. Lindigkeit, P. Langen, and J. Richter, eds.), pp. 231–235.

MAALØE, O., and KJELDGAARD, N. O. (1966). "Control of Macromolecular Synthesis." Benjamin, New York.

MACDONALD, R. E., TURNOCK, G., and FORCHHAMMER, J. (1967). The synthesis and function of ribosomes in a new mutant of *Escherichia coli*. *Proc. Natl. Acad. Sci. U.S.* **57,** 141–147.

MANOR, H., GOODMAN, D., and STENT, G. S. (1969). RNA chain growth rates in *Escherichia coli*. *J. Mol. Biol.* **39,** 1–27.

MOORE, P. B., TRAUT, R. R., NOLLER, H., PEARSON, P., and DELIUS, H. (1968). Ribosomal proteins of *Escherichia coli*. II. Proteins from the 30 S subunit. *J. Mol. Biol.* **31,** 441–461.

MORRIS, D. W., and KJELDGAARD, N. O. (1968). Evidence for the non-coordinate regulation of ribonucleic acid synthesis in stringent strains of *Escherichia coli*. *J. Mol. Biol.* **31,** 145–148.

NEIDHARDT, F. C. (1963). Properties of a bacterial mutant lacking amino acid control of RNA synthesis. *Biochim. Biophys. Acta* **68,** 365–379.

NOVICK, A., and WEINER, M. (1957). Enzyme induction as an all-or-none phenomenon. *Proc. Natl. Acad. Sci. U.S.* **43,** 553–566.

REVEL, M., LELONG, J. C., BRAWERMAN, G., and GROS, F. (1968). Function of three protein factors and ribosomal subunits in the initiation of protein synthesis in *E. coli. Nature* **219,** 1016–1021.

ROSSET, R., JULIEN, J., and MONIER, R. (1966). Ribonucleic acid composition of bacteria as a function of growth rate. *J. Mol. Biol.* **18,** 308–320.

SCHAECHTER, M., MAALØE, O., and KJELDGAARD, N. O. (1958). Dependency on medium and temperature of cell size and chemical composition during balanced growth of *Salmonella typhimurium. J. Gen. Microbiol.* **19,** 592–606.

SCHLEIF, R. F. (1967a). Control of ribosomal protein production in *E. coli.* Ph.D. Thesis, University of California, Berkeley.

SCHLEIF, R. F. (1967b). Control of production of ribosomal protein. *J. Mol. Biol.* **27,** 41–55.

SCHLEIF, R. F. (1968). Origin of chloramphenicol particle protein. *J. Mol. Biol.* **37,** 119–129.

SHIN, D. H., and MOLDAVE, K. (1966). Effect of ribosomes on the biosynthesis of ribonucleic acid *in vitro. J. Mol. Biol.* **21,** 231–245.

SILVERSTONE, A. E., MAGASANIK, B., REZNIKOFF, W. S., MILLER, J. H., and BECKWITH, J. R. (1969). Catabolite sensitive site of the lac operon. *Nature* **221,** 1012–1014.

STENT, G. S. (1964). The operon: On its third anniversary. *Science* **144,** 816–820.

STENT, G. S., and BRENNER, S. (1961). A genetic locus for the regulation of ribonucleic acid synthesis. *Proc. Natl. Acad. Sci. U.S.* **47,** 2005–2014.

STUBBS, J. D., and HALL, B. D. (1968). Effects of amino acid starvation upon constitutive tryptophan messenger RNA synthesis. *J. Mol. Biol.* **37,** 303–312.

YANOFSKY, C., and ITO, J. (1966). Nonsense codons and polarity in the tryptophan operon. *J. Mol. Biol.* **21,** 313–334.

DEVELOPMENTAL BIOLOGY SUPPLEMENT 3, 59–82 (1969)

Intracellular Communication in Early Animal Development

J. B. GURDON

Department of Zoology, Oxford University, Oxford, England

INTRODUCTION

The early development of animal eggs presents some of the clearest examples of intracellular communication. This is because, once fertilized, most animal eggs are effectively sealed off from their environment (e.g., by jelly coats in Amphibia, a chorion in insects, etc.). From fertilization until hatching, development is largely independent of the environment, and eggs will develop normally in the absence of light, nutrients, etc. Furthermore many different cell types appear during this period, and some of these are highly specialized, like muscle cells. This degree of cell differentiation and morphogenesis is therefore achieved by the interaction of components already present in the fertilized egg. There are two reasons why this interaction must involve communication between nucleus and cytoplasm. First, the organization of the egg cytoplasm is of critical importance for subsequent development, as shown by the harmful effect of experimental disturbances to it; second, nuclear activity is essential for postcleavage development as shown by the lethality of nucleocytoplasmic species hybrids and of doses of actinomycin D that suppress RNA synthesis.

The attention of this article is restricted to one kind of intracellular communication—that from the cytoplasm to the nucleus. Not only is communication of this kind thought to be very important in development, but the *nature* of communication in the other direction—the selection of gene products that are synthesized and that pass from the nucleus to the cytoplasm—is probably determined very largely by properties of the cytoplasm.

The aim of this article is (1) to summarize the evidence that cytoplasmic communication with the nucleus is important in normal development, (2) to review the nature of the molecules supposed to participate in communication, and (3) to suggest a way in which they may exert their effects. It is pointed out that the principles of intracellular communication thought to apply in early animal development may also apply to growing and dividing adult cells. However, all reference to intracellular communication in single-celled organisms is omitted, in order to permit a more detailed discussion of this phenomenon in the cells of multicellular organisms.

EVIDENCE FOR COMMUNICATION BETWEEN THE CYTOPLASM AND NUCLEUS OF DIFFERENTIATING CELLS

Specialized Regions of Egg Cytoplasm

The evidence that communication between cytoplasm and nucleus takes place in differentiating cells comes from several different experiments, all of which depend upon relating a certain kind of nuclear activity to the nature of its surrounding cytoplasm. In many cases it is possible to show a relationship between a particular region of egg cytoplasm and a certain type of cell differentiation, though it is not always certain that a change in nuclear activity is also involved. For example, the yellow cytoplasm of *Styela* is related to mesodermal differentiation (Conklin, 1905); in frogs the gray crescent region of the egg, formed at fertilization, is related to the site of the future blastopore lip and embryo axis (review by Pasteels, 1964). It is likely that nuclear activity is required for the realization of these cytoplasmic effects, but communication between cytoplasm and nucleus is demonstrated only if the activity of the nucleus is *changed* by the cytoplasm around it. The difficulty here is that local changes in nuclear activity are very hard to recognize biochemically, on account of the difficulty of collecting enough material from one part of an embryo. Advantage must therefore be taken of a few, rather unusual situations in which cytologically detectable nuclear changes take place in normal development.

Many animal eggs contain a cytochemically distinct "germ plasm" or "pole plasm." Some of the nuclei formed during cleavage become closely associated with, or surrounded by, this kind of cytoplasm, and it is these cells which, in later development, become the germ cells. In insects and Amphibia the germ plasm-associated nuclei at first divide more slowly than other nuclei at the same stage of development. In Amphibia, the germ plasm loses the special staining properties by which it is recognized at just the time when the germ cells start to proliferate in the gonad (Blackler, 1966). A more obvious effect of germ-line cytoplasm is to cause an apparent protection of associated nuclei from chromosome elimination. Proof that the nonelimination of chromosomes is really due to some property of the pole plasm has been obtained from experiments in which the pole plasm is irradiated with UV, or displaced by centrifugation (Fig. 1), or in which the egg is constricted so as to prevent any nuclei entering the pole plasm (Geyer-Duszyńska, 1959). The effect of such procedures is to prevent the normal association between pole cell nuclei and pole plasm; as a result the pole cell nuclei undergo chromosome elimination. Thus the cytoplasm clearly communicates with the nucleus in these circumstances, but it must be admitted that neither the germ plasm nor the pole plasm have yet been shown to alter the pattern of nuclear transcription, and in Amphibia, where chromosome elimination does not occur, the germ plasm has not yet been shown to influence nuclear activity in any other way than to reduce its rate of division.

One way of trying to relate a region of egg cytoplasm to an altered pattern of transcription is to remove part of the cytoplasm of an egg and to determine the effect of this on RNA synthesis. The polar lobe of several mollusc and annelid embryos is protruded from a blastomere before its division in such a way that the polar lobe material is finally restricted to cells which are essential for the mesendodermal differentiation of the embryo, as in *Ilyanassa* (Clement, 1952). Removal of the polar lobe leads to a quantitative reduction in the amount of RNA synthesized by the delobated embryos before they have come to differ in other respects from control embryos (Davidson *et al.*, 1965). It would be particularly interesting to know whether

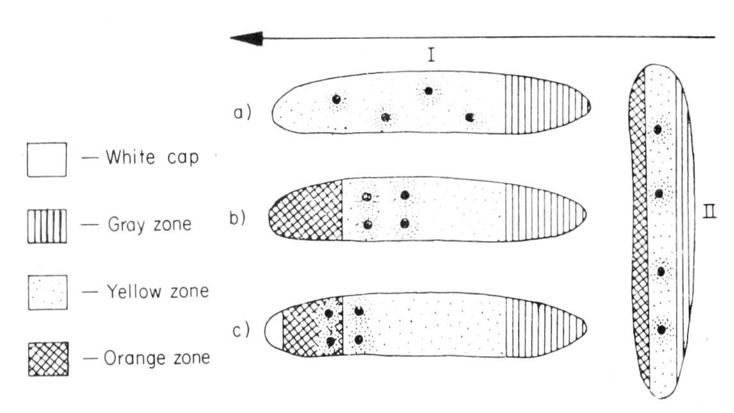

FIG. 1. Centrifugation of *Wachthiella* (Cecidomyiidae) eggs causes a rearrangement of different regions of cytoplasm and a displacement of cleavage nuclei. In spite of this, the pole plasm becomes associated with some cleavage nuclei which then fail to undergo chromosome elimination. From Geyer-Duszyńska (1959).

polar lobe cytoplasm has a qualitative as well as quantitative effect on nuclear RNA synthesis.

Hybrid Cultured Cells

A potentially informative experimental condition under which to study intracellular communication is in hybrid cultured cells (review by Ephrussi and Weiss, 1969). In some cases, stable lines of hybrid cells have been obtained, but for the purposes of analyzing nucleo-cytoplasmic interactions, most kinds of hybrid cells present a very complicated condition which is hard to interpret. They consist of a mixture of two kinds of nuclei and two kinds of cytoplasm. Although it may be assumed that the two kinds of cytoplasm are well mixed, it is entirely possible that each nucleus will respond primarily to its own cytoplasmic signals, which it may indeed continue to propagate, in the hybrid cell. Some interference in such overlapping regulatory cycles could lead to the altered gene expression often observed. The possible complexity of intracellular events in hybrid cells is much reduced in the case of the fusion of erythrocytes to other cells, since the erythrocyte cytoplasm is lost before fusion (Harris, 1967). It seems clear that any changes in the activity of the erythrocyte nucleus must depend on "signals" from the host cell cytoplasm or from its nucleus. The problem here is the significance of the imposed nuclear responses, since hybrid cells of this type do not survive to form growing lines of cells, and it could be argued that the type of intracellular communication revealed by these experiments is not necessarily the same as that observed in normal development or normal cell function. Nevertheless, many pronounced changes in nuclear activity take place in hybrid cultured cells, and such experiments have provided important evidence for the existence of intracellular communication.

Nuclear Transplantation Experiments

Nuclear transfer experiments in multicellular organisms, so far carried out mainly on Amphibia, offer several special advantages for the study of intracellular communication. First, the host cell (an egg or oocyte) can be enucleated, thereby ensuring that the nuclei subsequently transplanted are responding to the host-cell cytoplasm, not to factors emerging from the host cell nucleus. Second, the amount of cytoplasm introduced with the donor nucleus is trivial (1/50,000 or less) compared with that present in host cell. Third, and of most importance, many of the eggs receiving transplanted nuclei develop entirely normally. The reaction of transplanted nuclei to egg cytoplasm is therefore of a kind that takes place in *normal development*; it is not a pathological or unnatural response. In some nuclear transfer experiments normal development fails to take place. This happens when many nuclei are injected into one egg (as also when many sperm enter an egg of a species not naturally polyspermic) and in injected oocytes. Since, in such cases, nuclei change their activity so as to coincide with that characteristic of whatever host cell is used, it is assumed that the changes are of a nonpathological kind.

The results of these experiments have been described elsewhere (Gurdon and Woodland, 1968; Gurdon, 1968). Gross changes in nuclear activity can be rapidly induced in adult brain nuclei by injecting them into eggs (which induce DNA synthesis), into oocytes (which promote RNA synthesis), or into oocytes undergoing meiotic division. The latter suppress RNA and DNA synthesis and cause chromosome condensation as for nuclear division (Fig. 2). Another kind of nuclear transfer experiment which demonstrates an effect of egg components on nuclear function is summarized in Table I. Very shortly after the transfer of an embryo cell nucleus to egg cytoplasm, all RNA synthesis is suppressed. As the nuclear-transplant egg passes through the stages of early development, each main class of RNA is sequentially reactivated: nuclear, transfer, and ribosomal RNA in turn. Such experiments suggest, but do not prove, that these kinds of nuclear expression are controlled independently by different cytoplasmic components. Support for the independent control of these nuclear activities has come from nuclear transfer experiments between two genera of frogs (Woodland and Gurdon, 1969). It was found that the ratio of nuclear to transfer RNA synthesized by *Xenopus* nuclei in *Discoglossus* cytoplasm is 2.5 times lower than by *Xenopus* nuclei in *Xenopus* cytoplasm (Table II). The simplest explanation of this result is that *Xenopus* egg cytoplasm lacks components or conditions necessary for the activation of most *Discoglossus* "nuclear RNA" genes, but that transfer RNA genes are activated as usual by the foreign cytoplasm. It is clear from these experiments that living nuclei undergo changes in activity in response to components of living egg cytoplasm, and that communication of this kind normally takes place in development.

The demonstration that nuclear activity is controlled by cytoplasm

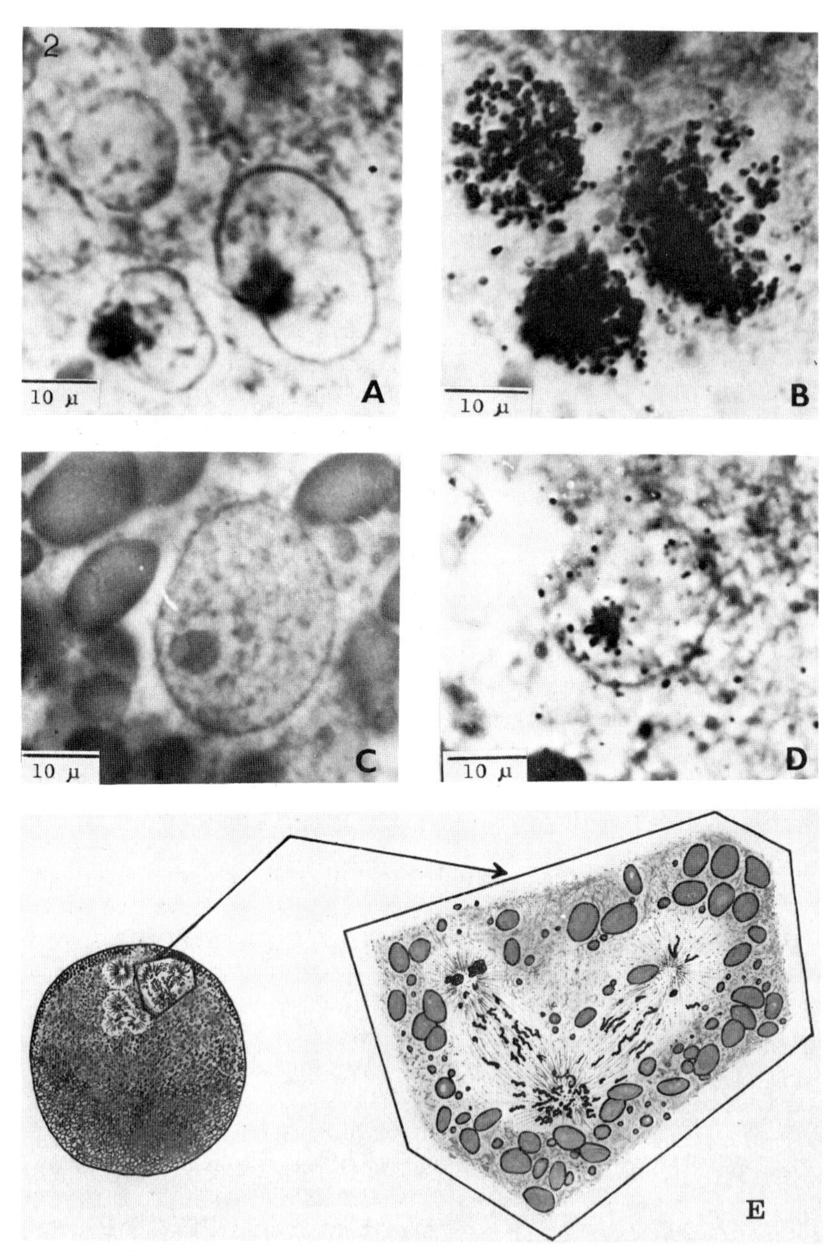

FIG. 2. The effects of cell cytoplasm on nuclear activity is shown by injecting adult frog brain nuclei into unfertilized egg cytoplasm (A), which induces DNA synthesis (autoradiography after TdR-^{3}H labeling-B), or into oocyte cytoplasm (C), which promotes RNA synthesis especially of the nucleolus (D, uridine-^{3}H-labeling). (E) Low and high power drawings of the condensed chromosomes that brain nuclei are induced to form after injection into oocytes undergoing completion of meiosis and conversion into eggs. Further details in Gurdon and Woodland (1968) and Gurdon (1968).

TABLE I
SUMMARY OF NUCLEAR TRANSFER EXPERIMENTS IN *Xenopus laevis* DEMONSTRATING AN INFLUENCE OF LIVING CYTOPLASM ON NUCLEAR ACTIVITY[a]

	Synthetic activity of embryos[b]			
	DNA	nRNA	tRNA	rRNA
Neurula cell (donor nuclei)	−	++	++	++
Nuclear-transplant embryos				
Uncleaved egg (1 hour after transfer)	++	−−	−−	−−
Mid blastula (7 hours after transfer)	+	++	−−	−−
Late blastula (9 hours after transfer)	+	++	++	−−
Neurula	−	++	++	++

[a] Single neurula nuclei were transplanted to enucleated eggs, which were labeled with uridine-^{3}H for 1–2 hours at various stages during their subsequent development. For details of experiments, see Gurdon and Woodland (1969).

[b] Symbols: −−, no detectable synthesis; −, c. 10% of nuclei active; +, c. 50% of nuclei active; ++, rapid synthesis in nearly all nuclei.

TABLE II
EFFECT OF FOREIGN CYTOPLASM ON RNA SYNTHESIS IN AMPHIBIAN NUCLEAR-TRANSPLANT HYBRIDS

Classes of RNA[a] (and method of fractionation): incorporation into 2 classes expressed as a ratio		Ratio of c.p.m. in RNA extracted from nuclear-transplant embryos[b]		Reduction in hybrids as % of control
		Control *Xen* → (*Xen*)	Hybrid *Disc* → (*Xen*)	
HMW RNA / 4s RNA	(Sephadex G-100)	1.2	0.51	42.5
rRNA / 4s RNA	(Sucrose gradients)	2.69	0.25[c]	9.5[c]
	(MAK columns)	2.32	0.22[c]	9.6[c]

SOURCE: Summarized from Woodland and Gurdon, 1969.

[a] HMW RNA = RNA excluded by G-100 Sephadex and shown by other means not to include ribosomal RNA; rRNA = 28 and 18s RNA on sucrose gradients, and RNA eluted by a high salt concentration from MAK columns; 4s RNA = RNA with properties of transfer RNA as judged by the procedures used.

[b] *Xen* = *Xenopus*, *Disc* = *Discoglossus*.

[c] These values are based on the assumption that all RNA sedimented or eluted in the position of rRNA is ribosomal RNA; it is doubtful if any of this RNA is, in fact, ribosomal in the hybrid embryos.

in *normal* cells is not restricted to oogenesis and the early stages of development. Several years ago Carlson (1952) performed what amounted to a nuclear-transplant experiment on grasshopper neuroblasts. These cells undergo several unequal divisions in which one daughter cell differentiates into a ganglion cell while the other becomes a neuroblast which repeats the process. The cytoplasm at the two sides of the dividing cell is visibly different. By means of a microneedle, Carlson was able to rotate the mitotic spindle within the neuroblast by about 180°, so that the daughter chromosomes which would have entered the part of the neuroblast cell destined to become a ganglion cell in fact went to the other side of the parent cell (Fig. 3). In spite of this the normal pattern of differentiation into neuroblast and ganglion cell took place, thereby demonstrating that ganglion cell differentiation depends on a property of the cytoplasm and not of the chromosomes. The pattern of unequal neuroblast division is very reminiscent of that observed in proliferating vertebrate epithelia, and it is a reasonable guess that the specialization

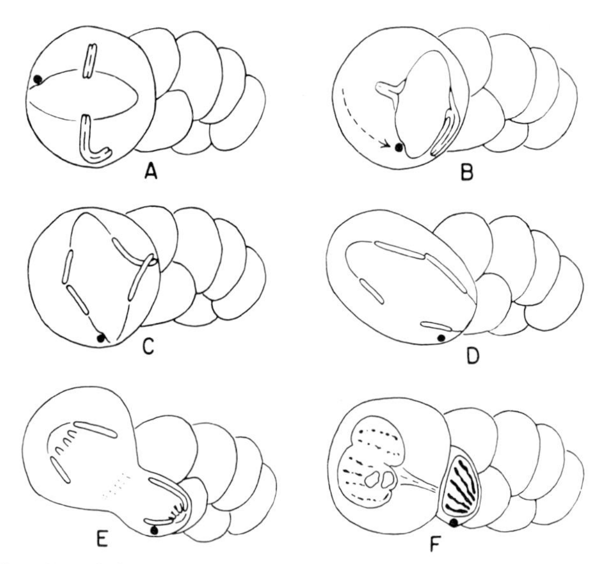

FIG. 3. Rotation of the spindle through 180°, using a microneedle (black spot), in a grasshopper neuroblast causes the chromosomes that would normally have entered the daughter neuroblast cell to enter the prospective ganglion cell. In spite of this, the usual unequal specialization of the daughter cells takes place, apparently through an effect of a distinct region of cytoplasm which enters the ganglion cell. From Carlson (1952).

of epithelial daughter cells may also depend on cytoplasmic composition.

THE NATURE OF CYTOPLASMIC COMPONENTS THAT INFLUENCE NUCLEAR ACTIVITY

Communication between the cytoplasm and nucleus of a cell is assumed, in the present discussion, to be achieved by the passage of molecules from the cytoplasm to the nucleus. Other means of communication (such as a loss of ions from the cytoplasm, leading to a loss of ions from the nucleus or chromosomes) are not known to be important in early development and are not considered here.

Four principal experimental approaches have been used to try and identify cytoplasmic molecules presumed to affect nuclear activity. The first is to describe differences between a kind of cytoplasm that has a certain effect and one that does not. Another approach is to disturb the organization or composition of cytoplasm and draw a conclusion about the effective component of the cytoplasm from the nature of the disturbing influence. A third line of investigation is to follow the movement of cytoplasmic molecules. These can be labeled and their passage into the nucleus related to changes in nuclear activity; alternatively, the molecular composition of nuclei can be compared before and after residence in a new cytoplasmic environment. Finally, attempts have been made to alter the activity of cells by incubating them in extracts of cytoplasm.

Pole Plasm and Germ Plasm

The properties of germ line and somatic cell cytoplasm have been compared in insect and amphibian embryos. Centrifugation causes displacement of the cytoplasmic component which prevents chromosome elimination in *Parascaris* (Boveri, 1910; Hogue, 1910). In Insects, granules can be displaced from the pole plasm by centrifugation and can be observed to associate with other nuclei in the middle of the embryo, which then fail to eliminate chromosomes (Fig. 1) (Geyer-Duszyńska, 1959). Electron microscopy shows the pole plasm of insects to contain aggregations of ribosome-sized particles embedded in a finely fibrous or granular material (Mahowald, 1962; Ullmann, 1965). The effective cytoplasmic component of germ-line cytoplasm seems to contain RNA or DNA because it is very sensitive to UV inactivation in insects (Geyer-Duszyńska, 1959) and in *Ascaris* (Moritz, 1967). The germ plasm of Amphibia (which has not been proved to affect nuclear expression) shows the spectrum of UV sensitivity at different wavelengths expected of nucleic acids but not of protein (Smith, 1966).

Entry of Cytoplasmic Protein into Nuclei in Fused Cells and Nuclear-Transplant Eggs

A direct demonstration of cytoplasmic molecules entering nuclei has been provided by cell fusion and nuclear transfer experiments. One to two days after a hen erythrocyte has been fused with a HeLa cell, DNA and RNA synthesis is induced in the erythrocyte nucleus. Bolund *et al.* (1969) have shown by interference microscopy that the activated erythrocyte nuclei undergo, after fusion, a severalfold increase in dry mass, presumably due to the entry of cytoplasmic protein. It has not been possible to relate a certain kind of protein to any one of the many kinds of response shown by the erythrocyte nuclei.

The passage of cytoplasmic protein into nuclei has been observed by autoradiography in amphibian nuclear-transfer experiments making use of the capacity of frog egg cytoplasm to induce DNA synthesis in adult brain nuclei. Arms (1968) labeled the cytoplasm of fertilized eggs with ^{3}H-amino acids; 2 hours later, when nearly all labeled amino acids had been incorporated into protein, puromycin and brain nuclei were injected. Autoradiography of the sectioned eggs fixed 1.5 hours after nuclear injection showed that labeled protein was present in the nuclei at about twice the concentration of that in the cytoplasm. The dose of puromycin used was shown, in separate experiments, to repress protein synthesis almost completely. In similar experiments, Merriam (1969) injected adult brain nuclei into unfertilized eggs whose cytoplasmic protein had been labeled by supplying ^{3}H-amino acids during oogenesis. Brain nuclei which had enlarged, as do the majority, had concentrated the labeled protein within an hour of injection, but a few nuclei that failed to enlarge, also failed to accumulate the labeled protein. This was an important observation because previous work (Graham *et al.*, 1966) had shown that only those brain nuclei which enlarge respond to egg cytoplasm by synthesizing DNA. Since egg cytoplasm induces DNA synthesis but represses RNA synthesis (see above), a relationship is established in these experiments between the entry of cytoplasmic protein into nuclei, and a particular kind of changed activity.

DNA Polymerase Activity in Early Development

The variable size of yolk platelets makes it very hard, if not impossible, to reisolate injected nuclei and identify the molecules that they have acquired during residence in egg cytoplasm. However, from information obtained in other ways it seems very likely that DNA polymerase (and associated enzymes) are among the molecules which enter brain and gamete nuclei and which induce DNA synthesis. The first kind of evidence supporting this conclusion is that purified DNA serves as a template for replication when introduced into egg cytoplasm, but does not do so in oocyte cytoplasm (Fig. 4). This difference coincides with the observation that nuclei transplanted to egg cytoplasm are rapidly induced to synthesize DNA, but the same nuclei transplanted to oocyte cytoplasm are not. These DNA injection experiments provide a kind of assay for DNA polymerase activity in *living* cells, and the results may be compared with *in vitro*

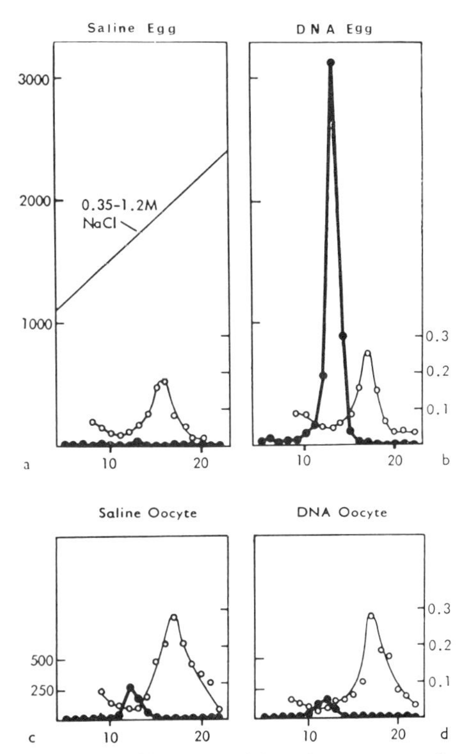

FIG. 4. Purified vertebrate DNA injected into frog egg cytoplasm serves as a template for DNA synthesis (a, b). In contrast, DNA injected into oocytes causes no stimulation of DNA synthesis (c, d). ○——○, OD_{260}, contributed mainly by egg or oocyte ribosomal RNA. ●——●, thymidine-^{3}H-labeled DNA. From Gurdon and Speight (1969).

assays for DNA polymerase, which have been carried out on eggs of sea urchins by Mazia and his colleagues but not as yet on frogs' eggs. Mazia and Hinegardner (1963) and Mazia (1966) have deduced from *in vitro* assays of DNA polymerase activity that DNA polymerase enters sperm nuclei after fertilization and is dissociated from mitotic chromosomes during cleavage. The enzyme(s) presumably becomes associated with the chromosomes during mitotic telophase because, in sea urchins and frogs, DNA synthesis takes place immediately after mitosis. These results suggest that DNA polymerase (and associated) molecules enter the cytoplasm of eggs during meiosis and become associated with the chromosomes of egg and sperm pronuclei, and then with those of cleavage nuclei, after each nuclear division. If this should turn out to be a correct interpretation, this situation would constitute the clearest example at present known of communication between the cytoplasm and nuclear chromosomes, involving a *known molecule* which causes a change in chromosome activity.

The Incubation of Cells in Cytoplasmic Extracts

The last and most direct approach to the identification of cytoplasmic molecules capable of affecting nuclear activity is to incubate cells or nuclei in cytoplasmic extracts. In our experience, isolated brain nuclei cannot be made to commence DNA synthesis by incubation with extracts of egg cytoplasm. Thompson and McCarthy (1968) have reported an effect of regenerating liver cytoplasm on isolated erythrocyte and liver nuclei. Although stimulation of both RNA and DNA synthesis is obtained, the total amount of synthesis is very small ($<10^{-6}$ of the amount of DNA in the nuclei), and the reaction is nearly complete within a few minutes. In view of these facts it is uncertain whether the stimulated incorporation which is observed really represents the normal process by which cell cytoplasm induces chromosome replication. The DNA-synthesis inducing factor is heat stable (therefore not DNA polymerase) and withstands freezing and thawing.

The incubation of whole cells with cytoplasmic extracts has given interesting results in at least two cases. Yamana and Shiokawa (1966) found some years ago that ribosomal RNA synthesis could be inhibited by up to 50% in cultured *Xenopus* neurula cells if they were grown (a) in a mixed culture with blastula cells, or (b) in the medium in which blastula cells had previously been grown (Fig. 5). Other experiments (references in Shiokawa and Yamana, 1969) show that their

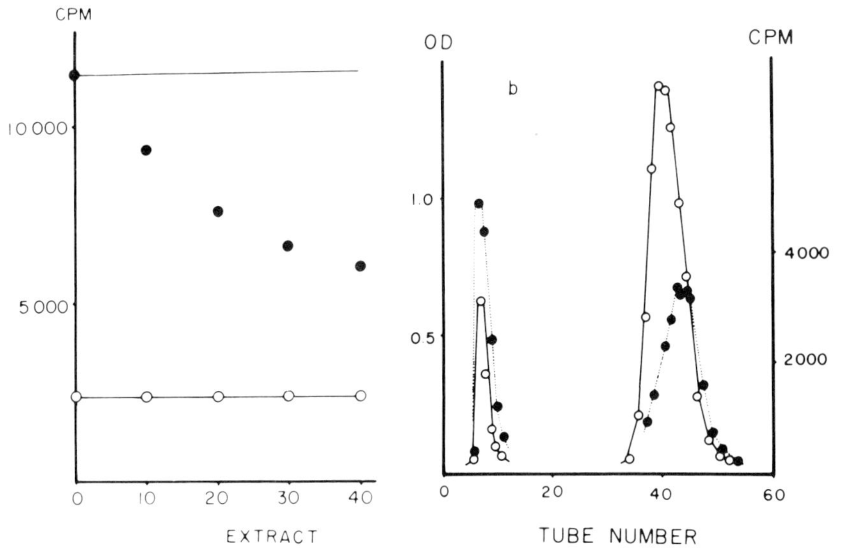

FIG. 5. The figure on the left shows the effect of increasing amounts of frog embryo cell extract on ribosomal RNA synthesis (●), and on 4 S RNA synthesis (○——○). The figure on the right shows an MAK chromatogram of RNA synthesized by cells incubated with inhibitor in the medium. Ribosomal RNA (right-hand peak) is sharply separated from 4 S RNA (left). ○——○, OD_{260}; ●------●, cpm. From Yamana and Shiokawa (1966).

inhibitor does not affect the synthesis of soluble RNA or heterogeneous RNA and that it can be extracted from the endoderm but not from other parts of neurula embryos, an interesting result in view of the fact that in normal development, ribosomal RNA synthesis is not detectable in the endoderm until after the neurula stage (Woodland and Gurdon, 1968). The identification of the factor has not been reported, but it is evidently a small molecule, since it is heat stable and dialyzable (references from Shiokawa and Yamana, 1969). Landesman and Gross (1968) were unable to repeat these experiments, but used a different method of RNA fractionation. The reason for this difference is not clear, but it is possible that detailed attention must be paid to the conditions of culture and to the preparation of the extract, as well as to recovery of RNA during extraction and fractionation.

An attempt to identify egg cytoplasmic components by their effect on differentiating cells has also been made on sea urchins by Hörstadius *et al.* (1967). Isolated animal or vegetal halves of 16–32-cell embryos were incubated in seawater containing fractions of the supernatant material of unfertilized eggs. A fraction eluted from Dowex columns by high pH caused an extension of ciliation in animal halves, and caused vegetal halves to develop nearly normally (Fig. 6). In contrast to these "animalizing" effects, another high pH fraction had a weak vegetalizing influence. It is not yet certain that these vegetalizing and animalizing materials (presumably small molecules) exert this effect through the nucleus, an assumption not required by their effects on development.

The little that is known about the nature of cytoplasmic molecules capable of altering nuclear activity can be summarized as follows. The major changes in nuclear expression induced by egg and cultured cell cytoplasm are accompanied by the entry of cytoplasmic protein into nuclei. An essential ingredient of pole plasm and germ plasm is RNA, but this is not known to alter nuclear DNA or RNA synthesis. A nonprotein, low molecular weight component of egg cytoplasm is able to repress ribosomal RNA synthesis by up to 50%, but

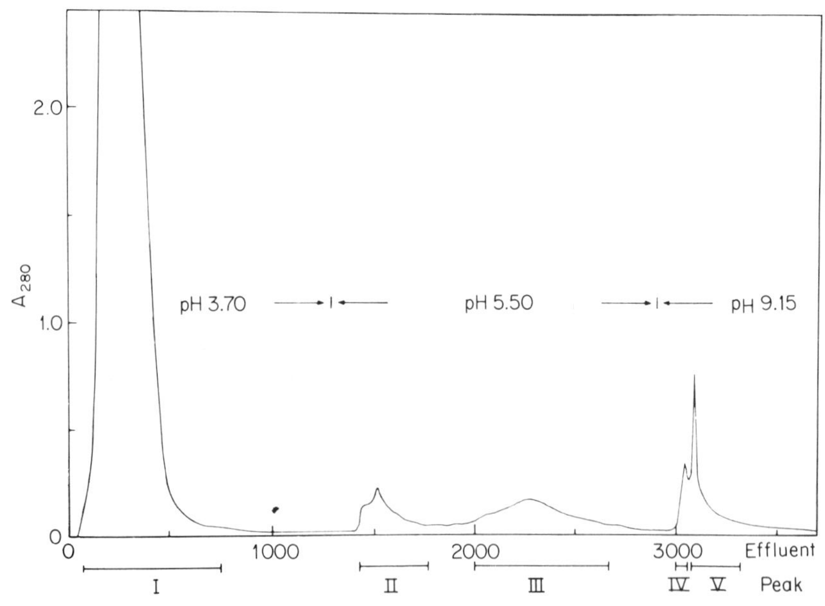

FIG. 6. Dowex chromatography of extracts from sea urchin eggs. Fractions IV and V cause animalization and vegetalization of half-embryos incubated in their presence. From Hörstadius *et al.* (1967).

may not be the only factor that controls ribosomal RNA synthesis. As far as the evidence goes, cytoplasmic proteins may be of particular importance in communicating with the nucleus.

THE PASSAGE OF CYTOPLASMIC MOLECULES INTO THE NUCLEUS, AND THEIR ASSOCIATION WITH CHROMOSOMES, IN NONDIFFERENTIATING CELLS

The examples of communication so far discussed concern experimental conditions where a nucleus responds to a new cytoplasmic environment in the course of development or cell differentiation. We now consider to what extent communication between cytoplasm and nucleus also takes place in dividing and growing cells not undergoing differentiation.

The Passage of Cytoplasmic Materials into Interphase Nuclei

Evidence for this comes primarily from oocytes which have the advantage of being large enough for micromanipulation, and from cultured cells in which the rapid exchange of metabolites permits an effective chase to be achieved after a pulse of label.

Work on the permeability of the nuclear membrane of living oocytes, up to 1960, has been fully reviewed by Mirsky and Osawa (1961). The most informative experiments are those of Harding and Feldherr (1959) and Feldherr and Feldherr (1960). Frog oocytes, injected with polyvinylpyrrolidone (molecular weight about 40,000), showed a rapid shrinkage of the nucleus, an effect not caused by sucrose or other small molecules. Fluorescein-labeled γ-globulin (165,000 molecular weight) failed to enter the nucleus of *Cecropia* oocytes within 10 minutes of injection though it spread throughout the cytoplasm within this time. These experiments indicate an impermeability of the nuclear membrane of the oocyte to molecules of 40,000 molecular weight or greater, over a short period of time. As the authors pointed out, penetration of large molecules into the nucleus over longer time periods is not excluded. Recently we have tested the permeability of frog oocyte nuclei by injecting oocyte proteins labeled with ^{125}I. The advantage of this procedure is that the proteins used are those that normally exist in the cell tested and that the label is covalently bonded to the aromatic moiety of tyrosine. Most of the proteins comprising the samples tested were of 50,000 or greater molecular weight, as judged by Sephadex filtration. Autoradiography of the sectioned injected oocytes (Fig. 7) showed that after 1 hour's incubation the nucleus was some four times more heavily

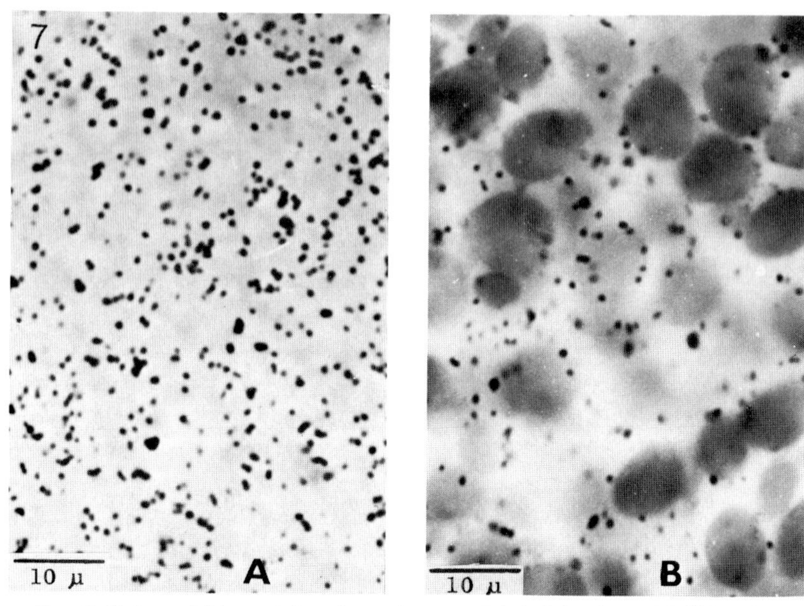

FIG. 7. Saline-soluble proteins of oocytes were labeled *in vitro* with ^{125}I and reinjected into living frog oocytes. The labeled molecules rapidly penetrate the nucleus, and the final ratio of nuclear (A) to cytoplasmic (B) grains is reached within 1 hour. (Weir and Gurdon, unpublished).

labeled than the cytoplasm, a value that did not change on further incubation for 24 hours (Weir and Gurdon, unpublished). It therefore seems that, in living frog oocytes, large cytoplasmic proteins enjoy relatively free access to the nucleus.

Using cultured L cells (mouse fibroblast line), Zetterberg (1966a, b) has demonstrated the passage of cytoplasmic protein into the nucleus by autoradiography and interference microscopy. After a 10-minute pulse of ^3H-labeled amino acids, cells were incubated in unlabeled amino acids. During the chase period the ratio of cytoplasmic to nuclear grains changed in favor of the nucleus, with little or no loss of material from the cell as a whole, until a constant ratio was reached 1–2 hours later. Although a protein synthesis inhibitor was not used, the extremely rapid exchange of free amino acids between these cells and the medium permits the conclusion that labeled protein has moved from the cytoplasm to the nucleus. In other experiments using interference microscopy (which measures dry mass, 80–90% of which is protein) it was shown that the total protein content of synchronized cells rises steadily during interphase. However, during the S phase, the nuclear content of protein increased by nearly two times, while the cytoplasm showed a slower increase at this stage of the cell cycle than during G_1. Four-minute pulses of leucine-^{3}H showed that the specific activity of the cytoplasmic protein (which reflects the rate of protein synthesis) was the same at all stages of the cell cycle. A flow of cytoplasmic protein evidently enters the nucleus during the S phase.

Two conclusions can be drawn from Zetterberg's experiments. First, cytoplasmic proteins exist that are synthesized in the cytoplasm but diffuse freely between the nucleus and cytoplasm of interphase cells. This conclusion has been confirmed on HeLa cells by Speer and Zimmerman (1968). After a brief labeling with leucine-^{3}H, cells were incubated in unlabeled leucine and cycloheximide. A progressive accumulation of label in the nucleus with an accompanying loss from the cytoplasm was observed, as expected if molecules synthesized in the cytoplasm gradually equilibrate with similar molecules in the nucleus. The work of Kroeger *et al.* (1963) on *Chironomus* salivary gland cells also indicates the passage of cytoplasmic proteins into interphase nuclei.

The second point demonstrated by Zetterberg's experiments is that a pronounced influx of cytoplasmic protein into the nucleus takes place during the S phase. Since it now seems clear that histones are synthesized in the cytoplasm during this phase (Robbins and Borun, 1967, who refer to earlier work), it seems likely that the proteins observed by Zetterberg to enter the nucleus at S phase may have been largely composed of histones.

The balance of results summarized above suggests that most cytoplasmic proteins diffuse into interphase nuclei. This does not, however, prove that they also become associated with chromosomes, though in the special case of histones it seems very likely that they become rapidly associated with newly synthesized DNA.

Exchange of Chromosome Materials during Mitosis

Mitosis seems to involve a rather complete dissociation of RNA from chromosomes in cultured cells (Prescott and Bender, 1963) and in amphibian embryos (Freedman *et al.*, 1967). The RNA associated with isolated mitotic chromosomes is mostly ribosomal and seems to be bound adventitiously during preparation (Salzman *et al.*, 1966).

Concerning basic proteins, Hancock (1969) has recently shown that in cultured HeLa cells the ratio of TdR-^{3}H-labeled DNA and lysine-^{14}C-labeled protein in isolated chromatin remains constant for at least eight cell cycles. This convincing result strongly indicates that histones are not dissociated from the DNA during cell division, a result consistent with the observation that about 50% of the protein associated with isolated interphase chromosomes is acid soluble (Maio and Schildkraut, 1967).

There appears to be little direct information about the behavior of chromosomal nonhistone proteins during mitosis. Prescott and Bender (1963) exposed cultured hamster fibroblasts to labeled amino acids, and determined the amount of labeled protein associated with metaphase chromosomes at each succeeding mitosis. The labeled protein was lost from the chromosomes much more rapidly than could be accounted for by chromosome dilution, the latter having been checked with thymidine-^{3}H-labeled DNA. The labeled amino acids used did not include arginine and only one was lysine; hence, it seems likely that the labeled protein studied was not histone. Using cytochemical measurements on dividing maize cells, Himes (1967) observed that the ratio of nonhistone protein to DNA was 50% lower in metaphase chromosomes than in interphase nuclei, although the ratio was the same for the genetically inert B chromosomes in the same nuclei. Richards (1960) observed by interference microscopy a rapid doubling of the protein content of prophase nuclei compared to mitotic chromosomes, over a period when this could not be accounted for by synthesis.

Two experiments have indicated the existence of a protein, associated with the nucleolus before division, that is dispersed from metaphase chromosomes, but returns to the newly formed nucleolus of the next interphase nucleus. This conclusion was reached by Das (1962) on the basis of the silver-staining affinity of nucleolar protein present in interphase nuclei of many different plant and animal cells. Using autoradiography of cultured rat connective tissue cells, Harris (1961) concluded that an valine-^{3}H-labeled protein, possibly of nucleolar origin, was incorporated into nucleoli after mitosis.

It is hard to draw any definite conclusion from these experiments. They do, however, seem to point to the possibility that nonhistone nuclear or chromosomal proteins are released from the nucleus or dissociated from chromosomes at mitosis, and return to the nucleus or chromosomes after mitosis.

THE MECHANISM OF CYTOPLASMIC COMMUNICATION WITH THE NUCLEUS

The following hypothetical scheme of communication is presented in order to bring together the rather heterogeneous range of observations just discussed. These may be summarized as follows. The existence of cytoplasmic communication with the nucleus seems to be very widespread, since an effect of cytoplasm on nuclear activity has been observed in every test so far carried out and in respect of all kinds of gene activity so far examined. Cytoplasmically induced changes in nuclear activity are associated with the passage of cytoplasmic protein into the nucleus. Some exchange of nuclear and cytoplasmic proteins takes place throughout interphase in non-differentiating cells, but DNA synthesis is accompanied by a pronounced influx of cytoplasmic protein. Nuclear, and probably non-histone chromosomal, proteins undergo a major exchange with cytoplasmic proteins during chromosome condensation at mitosis.

If, as suggested by these generalizations, gene activity is regulated by cytoplasmic proteins, there are two principal ways in which this could be achieved, according to whether the "regulatory" proteins have access to genes continuously, or only at certain stages of the cell cycle. Genes may also be classified into two categories, according to whether or not they are active for at least part of the cell cycle in all cells. Genes typically active in all cells include those that code for ribosomal RNA, transfer RNA, and for all enzymes and structural proteins associated with common cell organelles, like mitochondria. All, or nearly all, genes of single-celled organisms fall into this category. Other genes code for RNA's and proteins that are characteristic of certain specialized cell types, such as erythrocytes, muscle cells, etc., and these are usually assumed to be inactive in most cells. Activity of the first category of genes could be very simply controlled by a homeostatic system in which the activity of the gene is continuously regulated by the concentration of a direct or indirect product of the gene. A system of this kind would not account satisfactorily for the control of cell-type specific genes, since cells in which these genes are inactive (like most cells) are thought not to contain their products. Genes of this kind do not need to be subject to continual control, and a decision need be taken on their potential activity only once per cell cycle. This could be satisfactorily achieved by proteins that are synthesized in the cytoplasm and can gain access to the appropriate chromosome sites only at mitosis. If oocytes are typical of all cells in that cytoplasmic proteins enjoy free passage into the nucleus during interphase, it must be supposed that the restricted access of the cytoplasmic proteins to chromosomes is achieved by a displacement of other chromosomal proteins only once during the cell cycle. This could happen when chromosomes become condensed during mitosis; the displaced chromosomal proteins would be diluted out among cytoplasmic proteins, which could compete for the same chromosomal sites and which would have been synthesized in the cytoplasm during the immediately previous interphase period.

Although the relevant information presently available is not sufficiently precise to constitute a test of the scheme outlined, a number of experimental results are at least consistent with it.

At telophase of mitosis, a rapid dispersion of the condensed chromosome material takes place and the content of nuclear protein also increases rapidly. It has been suggested before (Gurdon and Woodland, 1968) that the pronounced nuclear swelling, chromosome dispersion, and ingress of cytoplasmic protein observed in nuclei transplanted to eggs may be functionally equivalent to the telophase reconstitution of nuclei in dividing cells. The justification for this view is that, at fertilization, the reconstitution of the egg pronucleus from condensed chromosomes and the enlargement of the interphase sperm nucleus are functionally equivalent, and the behavior of the sperm nucleus and transplanted nuclei are very similar. Our interpretation is that the enlargement of nuclei transplanted to eggs, and of erythrocyte nuclei in HeLa cell cytoplasm, as well as the normal reconstitution of nuclei after mitosis, are all occasions when cytoplasmic proteins reprogram the chromosomes for the activity or inactivity of cell-type specific genes during the next interphase period, as shown diagrammatically in Fig. 8.

If the reprogramming of cell-type specific genes is restricted to mitosis, we would expect major changes in cell differentiation to be associated with, or preceded by, cell division. Many situations are known where such a relationship exists (reviews by Ebert and Kaighn, 1967; Holtzer, 1968; Wessells, 1968), although it has not yet been proved that the reason for this relationship is connected with chromosome reprogramming.

A scheme of control, like that outlined above, would fit in with what is known of development most simply, if it is assumed that the postulated regulatory cytoplasmic proteins repress some genes,

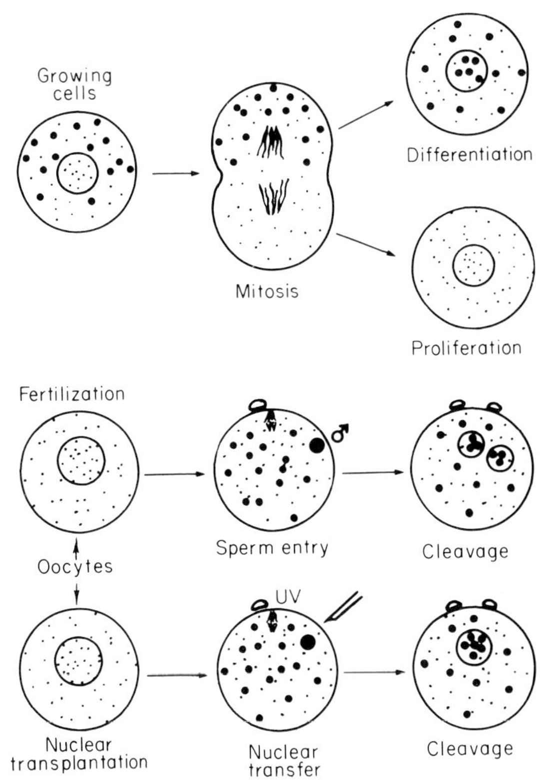

FIG. 8. Diagrams showing the inclusion of cytoplasmic molecules in nuclei. This happens during postmitotic nuclear reconstitution in growing cells (upper diagram) and in the egg pronucleus at fertilization (middle picture); it also happens during the massive swelling of the sperm nucleus at fertilization (middle diagram) and of a transplanted nucleus (lower diagram). In each case an unequal cell division is shown and the cytoplasmic molecules supposed to influence nuclear activity (large dots) are arranged so as to be included primarily in one of the two daughter cells and nuclei. The unequal division of the maturing oocyte into an egg and polar bodies is represented by the small objects shown above eggs.

but permit the activity of others needed to keep the cell and its daughter cells supplied with the same repressors. Thus each daughter cell would be reprogrammed at mitosis like its parent cell until developmentally important agents, such as unequally distributed components of egg cytoplasm, hormones, inducers, etc., should interfere with these self-reproducing cycles (Fig. 8). Schemes not unlike that described here have been postulated before, and interest in these proposals would mainly arise if further experiments along the lines of some of those discussed should add support to the concept of chromosome reprogramming at mitosis. Such a process is likely to be characteristic of multicellular organisms; if it exists in Bacteria at all, it should affect only those genes involved in events, like sporulation, that do not take place in every cell generation.

ACKNOWLEDGMENTS

The author's work referred to in this article was supported by the Medical Research Council of Great Britain, and was immensely facilitated by the very able assistance of Miss V. Speight.

REFERENCES

ARMS, K. (1968). Cytonucleoproteins in cleaving eggs of *Xenopus laevis*. *J. Embryol. Exptl. Morphol.* **20,** 367–374.

BLACKLER, A. W. (1966). Embryonic sex cells of amphibia. *Advan. Reprod. Physiol.* **1,** 1–28.

BOLUND, L., RINGERTZ, N. R., and HARRIS, H. (1969). Changes in the cytochemical properties of erythrocyte nuclei reactivated by cell fusion. *J. Cell Sci.* **4,** 71–88.

BOVERI, T. (1910). Ueber die Teilung zentrifugierter Eier von *Ascaris Megalocephala*. *Arch. Entwicklungsmech. Organ.* **30,** 101–125.

CARLSON, J. G. (1952). Microdissection studies of the dividing neuroblast of the grasshopper, *Chortophaga viridifasciata* (De Geer). *Chromosoma* **5,** 199–220.

CLEMENT, A. C. (1952). Experimental studies on germinal localization in *Ilyanassa*. I. Role of the polar lobe in determination of the cleavage pattern and its influence in later development. *J. Exptl. Zool.* **121,** 593–625.

CONKLIN, E. G. (1905). Organization and cell lineage of the ascidian egg. *J. Acad. Natl. Sci. Philadelphia* **13,** 1–119.

DAS, N. K. (1962). Demonstration of a non-RNA nucleolar fraction by silver staining. *Exptl. Cell Res.* **26,** 428–431.

DAVIDSON, E. H., HASLETT, G. W., FINNEY, R. J., ALLFREY, V. G., and MIRSKY, A. E. (1965). Evidence for prelocalisation of cytoplasmic factors affecting gene activation in early embryogenesis. *Proc. Natl. Acad. Sci. U.S.* **54,** 696–704.

EBERT, J. D., and KAIGHN, M. E. (1966). Keys to change: factors regulating differentiation. *In* "Major Problems in Developmental Biology" (M. Locke, ed.), pp. 29–84. Academic Press, New York.

EPHRUSSI, B., and WEISS, M. C. (1969). Hybrid somatic cells. *Sci. Am.* **220,** 26–35.

FELDHERR, C. M., and FELDHERR, A. B. (1960). The nuclear membrane as a barrier to the free diffusion of proteins. *Nature* **185,** 250–251.

FREEDMAN, M. L., STAMBROOK, P. J., and FLICKINGER, R. A. (1967). The absence of labeled RNA on metaphase chromosomes of *Taricha* and *Rana* embryos. *Exptl. Cell Res.* **47,** 640–643.

GEYER-DUSZYŃSKA, I. (1959). Experimental research on chromosome diminution in *Cecidomyiidae* (Diptera). *J. Exptl. Zool.* **141,** 391–441.

GRAHAM, C. F., ARMS, K., and GURDON, J. B. (1966). The induction of DNA synthesis by frog egg cytoplasm. *Develop. Biol.* **14,** 349–381.

GURDON, J. B. (1968). Changes in somatic cell nuclei inserted into growing and maturing amphibian oocytes. *J. Embryol. Exptl. Morphol.* **20,** 401–414.
GURDON, J. B. (1969). Nucleo-cytoplasmic interactions during cell differentiation. *Genetics* in press.
GURDON, J. B., and SPEIGHT, V. A. (1969). The appearance of cytoplasmic DNA polymerase activity during the maturation of amphibian oocytes into eggs. *Exptl. Cell Res.* **55,** 253–256.
GURDON, J. B., and WOODLAND, H. R. (1968). The cytoplasmic control of nuclear activity in animal development. *Biol. Rev. Cambridge Phil. Soc.* **43,** 233–267.
GURDON, J. B., and WOODLAND, H. R. (1969). The influence of the cytoplasm on the nucleus during cell differentiation, with special reference to RNA synthesis during amphibian cleavage. *Proc. Roy. Soc.* **B173,** 99–111.
HANCOCK, R. (1969). Conservation of histones in chromatin during growth and mitosis *in vitro*. *J. Mol. Biol.* **40,** 457–466.
HARDING, G. V., and FELDHERR, C. M. (1959). Semipermeability of the nuclear membrane in the intact cell. *J. Gen. Physiol.* **42,** 1155–1165.
HARRIS, H. (1961). Formation of the nucleolus in animal cells. *Nature* **190,** 1077–1078.
HARRIS, H. (1967). The reactivation of the red cell nucleus. *J. Cell Sci.* **2,** 23–32.
HIMES, M. (1967). An analysis of heterochromatin in maize root tips. *J. Cell Biol.* **35,** 175–181.
HOGUE, M. J. (1910). Über die Wirkung der Zentrifugalkraft auf die Eier von *Ascaris megalocephala*. *Arch. Entwicklungsmech. Organ.* **29,** 109–145.
HOLTZER, H. (1968). Induction of chondrogenesis: a concept in quest of mechanisms. *In* "Epithelial-mesenchymal Interactions" (R. Fleischmajer and R. E. Billingham, eds.), pp. 152–164. Williams & Wilkins, Baltimore, Maryland.
HÖRSTADIUS, S., JOSEFSSON, L., and RUNNSTRÖM, J. (1967). Morphogenetic agents from unfertilized eggs of the sea urchin *Paracentrotus lividus*. *Develop. Biol.* **16,** 189–202.
KROEGER, H., JACOB, J., and SIRLIN, J. L. (1963). The movement of nuclear protein from the cytoplasm to the nucleus of salivary gland cells. *Exptl. Cell Res.* **31,** 416–423.
LANDESMAN, R., and GROSS, P. R. (1968). Patterns of macromolecule synthesis during development of *Xenopus laevis*. I. Incorporation of radioactive precursors into dissociated embryos. *Develop. Biol.* **18,** 571–589.
MAHOWALD, A. P. (1962). Fine structure of pole cells and polar granules in *Drosophila melanogaster*. *J. Exptl. Zool.* **151,** 201–215.
MAIO, J. J., and SCHILDKRAUT, C. L. (1967). Isolated mammalian metaphase chromosomes. I. General characteristics of nucleic acids and proteins. *J. Mol. Biol.* **24,** 29–39.
MAZIA, D. (1966). Biochemical aspects of mitosis. "The Cell Nucleus, Metabolism and Radiosensitivity," p. 15. Taylor & Francis, London.
MAZIA, D., and HINEGARDNER, R. T. (1963). Enzymes of DNA synthesis in nuclei of sea urchin embryos. *Proc. Natl. Acad. Sci. U.S.* **50,** 148–156.
MERRIAM, R. W. (1969). Movement of cytoplasmic proteins in nuclei induced to enlarge and initiate DNA or RNA synthesis. *J. Cell Sci.* **5,** in press.
MIRSKY, A. E., and OSAWA, S. (1961). *In* "The Cell" (J. Brachet and A. E. Mirsky, eds.), Vol. II, pp. 677–770. Academic Press, New York.
MORITZ, K. B. (1967). Die Blastomerendifferenzierung für Soma und Keimbahn bei

Parascaris equorum. II. Untersuchungen mittels UV-Bestrahlung und Zentrifugierung. *Arch. Entwicklungsmech. Organ.* **159,** 203–266.

Pasteels, J. J. (1964). The morphogenetic role of the cortex of the Amphibian egg. *Advan. Morphogenesis* **3,** 363–388.

Prescott, D. M., and Bender, M. A. (1963). Synthesis and behavior of nuclear proteins during the cell life cycle. *J. Cellular Comp. Physiol.* **62** Suppl., 175–194.

Richards, B. M. (1960). Redistribution of nuclear proteins during mitosis. *In* "The Cell Nucleus" (J. S. Mitchell, ed.), pp. 138–140. Butterworths, London.

Robbins, E., and Borun, T. W. (1967). The cytoplasmic synthesis of histones in HeLa cells and its temporal relationship to DNA replication. *Proc. Natl. Acad. Sci. U.S.* **57,** 409–416.

Salzman, N. P., Moore, D. E., and Mendelsohn, J. (1966). Isolation and characterization of human metaphase chromosomes. *Proc. Natl. Acad. Sci. U.S.* **56,** 1449–1456.

Shiokawa, K., and Yamana, K. N. (1969). Inhibitor or ribosomal RNA synthesis in *Xenopus laevis* embryos. II. Effects on ribosomal RNA synthesis in isolated cells from *Rana japonica* embryos. *Exptl. Cell Res.* **55,** 155–160.

Smith, L. D. (1966). The role of a "germinal plasm" in the formation of primordial germ cells in *Rana pipiens*. *Develop. Biol.* **14,** 330–347.

Speer, H. L., and Zimmerman, E. F. (1968). The transfer of proteins from cytoplasm to nucleus in HeLa cells. *Biochem. Biophys. Res. Commun.* **32,** 60–65.

Thompson, L. R., and McCarthy, B. J. (1968). Stimulation of nuclear DNA and RNA synthesis by cytoplasmic extracts *in vitro*. *Biochem. Biophys. Res. Commun.* **30,** 166–172.

Ullmann, S. L. (1965). Epsilon granules in *Drosophila* pole cells and oocytes. *J. Embryol. Exptl. Morphol.* **13,** 73–81.

Wessells, N. K. (1968). Problems in the analysis of determination, mitosis, and differentiation. *In* "Epithelial-mesenchymal Interactions" (R. Fleischmajer and R. E. Billingham, eds.), pp. 132–151. Williams & Wilkins, Baltimore, Maryland.

Woodland, H. R., and Gurdon, J. B. (1968). The relative rates of synthesis of DNA, sRNA and rRNA in the endodermal region and other parts of *Xenopus laevis* embryos. *J. Embryol. Exptl. Morphol.* **19,** 363–385.

Woodland, H. R., and Gurdon, J. B. (1969). RNA synthesis in an amphibian nuclear-transplant hybrid. *Develop. Biol.* **20,** 89–104.

Yamana, K., and Shiokawa, K. (1966). Ribonucleic acid (RNA) synthesis in dissociated embryonic cells of *Xenopus laevis*. II. Inhibitor of ribosomal RNA synthesis. *Proc. Japan. Acad.* **42,** 811–815.

Zetterberg, A. (1966a). Synthesis and accumulation of nuclear and cytoplasmic proteins during interphase in mouse fibroblasts *in vitro*. *Exptl. Cell Res.* **42,** 500–511.

Zetterberg, A. (1966b). Protein migration between cytoplasm and cell nucleus during interphase in mouse fibroblasts *in vitro*. *Exptl. Cell Res.* **43,** 526–536.

DEVELOPMENTAL BIOLOGY SUPPLEMENT 3, 83–111 (1969)

On the Centripetal Course of Development, the *Fucus* Egg, and Self-electrophoresis

LIONEL F. JAFFE

Department of Biological Sciences, Purdue University, West Lafayette, Indiana

A VIEWPOINT OF DEVELOPMENT AND COMMUNICATION

Development from the Inside Out, or from the Outside In?

I welcome this opportunity to discuss our work on developmental localization under the rubric of communication. For under this heading one can recommend an infusion of physiological concepts into developmental biology; in particular, concepts from the general physiology of communication, whether this be sensory, neural, or hormonal.

For too long, too many developmental biologists seem to have ingested too little other than *Escherichia coli* genetics. So pervasive and unchallenged has been the influence of bacterial genetics that some authors now actually *define* development as sequential protein synthesis. So imminent has the successful application of the Jacob–Monod model seemed, that there has been an extraordinary rush to support rather than truly to test it. Too often, almost any indicator that the genome, or RNA synthesis, or protein synthesis, is ultimately necessary for development or ultimately responds to some developmental signal is somehow taken as strong evidence for direct control of or direct control by the genome.

This dogma seems to rest upon a complex of five interlocking ideas. Let us briefly consider them in the dual light of communications physiology and developmental studies proper:

First of all this coliform vision seems to be essentially space free. Somehow, only the temporal dimension, only changes are truly focused upon. Spatial development is somehow considered to be secondary to and essentially similar to temporal development. Indeed, changes in time and the emergence of differences in space seem hardly to be distinguished so that the vague term differentiation is applied to both. Now it seems to me that the whole history of developmental biology shows *localization*, i.e., the emergence of pattern, to be the central problem of development. What, after all, does the truth of epigenesis as opposed to preformationism mean? In modern terms, it means that genetic *instructions* are converted into a developmental *map*. The architect's *words* are converted into a *blueprint*. This essential process of localization absolutely requires complex communication between the emerging and differentiating parts. That is the main reason why our knowledge of communication physiology should be so applicable to development. It is also true, of course, that the proper timing of development usually requires considerable input from the environment.

A second distortion of this vision is that the units of differentiation are necessarily cells. Should we disregard the marvelous patterns found within acellular or even uninucleate adults as of the ciliates? or the fine mosaic of determination often found in early acellular or even uninucleate stages of forms like the insects? or the considerable localization within such mature cells as giant nerve cells and polarized epithelia? Are these mere oddities, or are they only the more obvious indicators of a more general condition? Perhaps the recently renewed studies of the *connectedness* of various metazoan cells (e.g., those of Loewenstein, 1968; Furshpan and Potter, 1968) will better define the effective units of differentiation.

A third, most dubious assumption is that the key molecular events of development are total syntheses of macromolecules, particularly of proteins. Communications physiology should make us very wary of this protein synthesis dogma. The molecular events in neural communication that are presently known do not include any protein syntheses as direct links; in fact, the main events are transport processes particularly of ions and of transmitters. The best-analyzed hormonal action, that of epinephrine on liver cells, seems to start with some rapid membrane rearrangement and then goes through a long chain of molecular events each involving the alteration, never the synthesis, of a protein or small molecule (Sutherland *et al.*, 1965). Finally the only important molecular process so far known in the tropisms is that of auxin *transport* (Briggs, 1963).

A fourth notion is that of development via highly specific signals that can yield but one output, that determine much of that output's character, and that are closely linked to that output—specific gene repressors, for example. Again the available facts suggest that such signals are rare. It is common knowledge that few sensory inputs affecting behavior have this character. The best understood developmental signals are those in the vascular plants: if there is one thing abundantly clear about the main signals, blue and red light, as well as the five major hormones, it is the wide variety of outputs each can elicit.

The fifth and most pervasive of these genetic notions seems to be the assumption that because inheritance starts from the inside, from the DNA, development, too, must proceed centrifugally. However, the direct study of embryos, from those of Conklin and Morgan down to the recent marvels of nuclear transplants, have, of course, long pointed exactly the other way, that is to development from the outside in. Here I would like to develop the theme that our more general knowledge of communication physiology also points to the cell surface as the first, and the nucleus as the last, site of developmental change.

The Cell Surface in Development: Environmental Signals

Evidence now multiplies that it is the cell surface which is usually the target, and often the transmission line, signal generator, and even the memory bank in communication.

Thus the cell surface is proving to be the main target even of light signals. This finding is of particular weight because light is one of the few signals that *could* pass freely through the cell surface. The evidence lies mainly in the dependence of the polarization of the cell upon the polarization of the light in responses to both blue light (Jaffe and Etzold, 1962; Haupt, 1965), and red light (Jaffe and Etzold, 1965); of chloroplast movements in response to both blue and red light (Haupt, 1966), and even of vision.

In this last case, the recent elegant study of Waterman *et al.* (1969) yields a more direct indicator that the locus of the receptor molecules is the surface. In the case of cellular photopolarization the best evidence comes from a study of developing *Botrytis* spores (Jaffe and Etzold, 1962; see Fig. 1). In the closely related case of cellular phototropism, the only case so studied has been that of the *Phycomyces* sporangiophore. Unfortunately, geometrical reasons make the argument from polarization dependence quite indirect in this cylindrical cell (Jaffe, 1960). However, even a determined critic is now inclined to accept this argument (Delbrück, 1969).

An interesting by-product of the analysis of polarization dependence is the inference that in almost all cases the receptor molecule is excited by the electric vector in the light (Jaffe, 1962). However, when spores of the moss *Funaria* are polarized by very intense red light they grow out in the direction of the magnetic vector. It is

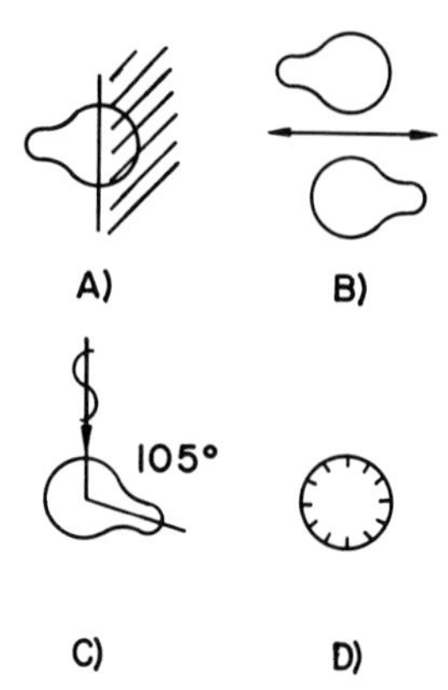

FIG. 1. Photoreceptor locus in *Botrytis* spores. If half illuminated with unpolarized light, the cells grow from their lit part (A); if lit from above and below with polarized light they grow out in the direction of the E-vector (B); if lit from the side with horizontally polarized light, they grow out at about 105° to the light (C). Careful analysis of the optics of the cell indicates that the receptor molecules must be in or near the plasma membrane and normal to it to explain these results (D). From Jaffe and Etzold (1962).

quite difficult to explain this finding without the remarkable inference that the receptor molecules for *this* response are being excited by the magnetic vector of the light (Jaffe and Etzold, 1965).

It has long been known that a very different signal, namely the *sperm*, must hit the outside of the egg in order to convey its message; artificially injected sperm fail to activate eggs. Moreover, the fact that excessive constriction can restrict activation to one end of an egg indicates that the cell surface is a transmission line as well as target for this signal (Monroy, 1965). A remarkable newer indicator of the role of the cell surface in fertilization is Russian evidence that an action potential communicates the fact of pollination from the stigma to the ovary (Sinyukhin and Britikov, 1967).

Finally, with regard to environmental inputs, it is of considerable interest to reconsider the response of bacteria to soluble nutrients. After all, this is the phenomenon the study of which generated the Jacob-Monod model. In eliciting a digestive enzyme and a "permease," the nutrient molecule goes nearly straight to the DNA (Jacob and Monod, 1963). But it now appears that in eliciting another response, namely movement of the bacteria toward regions of more concentrated food, the food molecule's target lies on the cell membrane's exterior. This seems to follow from the interesting, recent observation that mutants defective in the "permease" for a nutrient respond chemotactically to *lower* concentrations of this nutrient than do wild-type cells (Adler, 1969).

The Cell Surface in Development: Hormones and Other Signals

It is very well known that the cell surface is the target, transmission line, and signal generator in neural communication. Now, there is increasing evidence that this surface is likewise the usual target in hormonal communication.

Epinephrine and estrogen can act in seconds to somehow activate surface-bound adenyl cyclase. Thus in isolated rat hearts stimulated with physiological concentrations of epinephrine, cyclic 3′,5′-AMP rises by severalfold, reaching a peak in *2–4 seconds* after the addition of epinephrine to the perfusator (Sutherland *et al.*, 1965), whereas the cyclic AMP in the uteri of ovariectomized rats doubles, to reach its peak and normal level within *15 seconds* after intravenous injection of physiological doses of estradiol-17β (Szego and Davis, 1967). The very speed of these responses argues strongly for a membrane target. So do fractionation studies. Thus, certain nucleus-free red cell membrane preparations prove to contain an epinephrine activable cyclase (Sutherland *et al.*, 1965). Evidence for action via surface-bound cyclase is rapidly accumulating for many other vertebrate hormones, particularly peptide hormones such as ACTH, vasopressin, etc. (Robison, 1969).

In the insect *Galleria*, addition of a juvenile hormone preparation of physiological potency to isolated salivary glands, is reported to lower their membrane potential by as much as 35 mV within 1 minute. Moreover, the sodium dependence of the response suggests that it mainly indicates a rise in sodium conductance (Baumann, 1968).

In flowering plants there is evidence that addition of physiological concentrations of auxin can act in about 1 minute to cause a transient rise in the membrane potential of root cells (Jenkinson, 1962; Jenkinson and Scott, 1961) and can substitute for pollen in eliciting the action potential reported to activate the ovary (Sinyhukin and Britikov, 1967).

Better documented, if less direct, indicators that auxin has a surface target are its well known, rapid effects upon protoplasmic streaming, which are elicited in 20 seconds or less (Sweeney, 1941); recent evidence that even certain growth changes (the initial *inhibi-*

tion of coleoptile growth) may be elicited in 2 minutes or less (Evans, Rayle, and Hertel, unpublished); that polar transport and growth effects show the same dependence upon the structure of the auxin molecule (Hertel *et al.*, 1969); and finally that the kinetics of the growth response when analyzed in detail, seem incompatible with a genetic target (Ray, this symposium).

A remarkable, recent investigation of Smith and Ecker (1969) offers particularly cogent evidence for the obverse, evidence that the nucleus is *not* the target of hormonal communication. These authors prove that progesterone can induce the isolated and *enucleated* frog oocyte to mature (Fig. 2). Their further observation that even a 1- to 2-minute exposure to high concentrations (10 μg/ml) of progesterone suffices to induce ripening of oocytes again suggests that the hormone's target is in fact the cell surface. Similar, if less clear evi-

Fig. 2. Photomicrograph of enucleated and then progesterone-treated frog oocytes about 3 hours after artificial activation. The germinal vesicles were removed several hours before progesterone treatment. The oocytes then matured; 2 days after hormone treatment they were activated by being pricked with a clean needle. Every egg in the figure exhibits an abortive cleavage furrow. (From Smith and Ecker, 1969). Under certain conditions, enucleated and then hormone-treated oocytes will undergo movements that significantly resemble those of gastrulation (Smith and Ecker, unpublished data): so-called pseudogastrulation (Holtfreter, 1943).

dences of hormone effects in the absence of nuclear action have been reported for both insulin and vasopressin (Rodbell *et al.*, 1968; Edelman et al., 1963).

Many developmental signals of course *are* particular surfaces, both cellular and noncellular. For example, an air–water interface is needed to initiate localization in the *Dictyostelium* aggregate (Gerisch, 1968); follicle cells spiralize the *Limnaea* egg (Raven, 1967); normal vertebrate inductions usually seem to require cell contact; morphogenetic movements in vertebrates seem to be generally guided by cell contacts (Steinberg, 1964).

Reconsideration of the transfilter induction experiments suggests that they are better explained through filopods than through diffusible messengers. In the best-studied case, the inductive influence traversed pores of 25 $\mu \times 0.5\ \mu$, but was markedly impeded by pores of 70 $\mu \times 0.5\ \mu$ or of 25 $\mu \times 0.1\ \mu$ (Saxén *et al.*, 1968 p. 253). Considering that whole leukocytes can move through filters 150 μ thick with pores as small as 0.7 μ (Ward *et al.*, 1965), there seems to be no difficulty in imagining that communication occurred via filopods. Diffusible agents, however, would seem to have needed the implausible diameter of nearly 0.1 μ to explain these data.

Recent studies of one very important developmental signal, namely phytochrome-absorbed red light, indicate not only that the receptor is in the cell surface, but that its excitation involves a 90° rotation of the chromophore within this same surface (Etzold, 1965; Haupt, 1968); that its excitation may then produce a second signal which spreads in seconds and at rates suggesting a membrane-relayed wave (Wagné, 1965); that the first measurable, cellular, but still reversible consequences are surface changes (both adhesiveness and potential changes occur in 30 seconds or less and indicate the secretion of some positively charged component) (Tanada, 1968; M. J. Jaffe, 1968); and finally that of the two fastest known complete responses to this signal, namely chloroplast rotation and the leaf movements of certain plants, at least the latter are actually affected by a surface motor, namely, by the osmotic consequences of transmembrane ion transport (Hendricks and Borthwick, 1967).

There is, of course, abundant evidence that the *irreversible*, but still largely invisible consequences of a developmental signal, i.e., *determination*, lie in large part in or near the cell surface. One can cite to this point Curtis's transplants of the gray crescent's cortex (Curtis, 1963), the so-called subcortical accumulation, which indicate the spiralization of the *Limnaea* egg (Raven, 1967); the centrolecithal structure of an insect egg in a mosaic state of determination; the well known reestablishment of a normal arrangement in many stratified eggs; and the sorting out, presumably through differential adhesiveness of various determined but still undifferentiated cells, e.g., imaginal disc cells (García-Bellido, 1966).

Finally, at least in ciliates, even the *inheritance* of developmental patterns can be directly cortical (Tartar, 1962; Beisson and Sonneborn, 1965).

LOCALIZATION IN THE DEVELOPING *FUCUS* EGG

The early development of the Fucales egg is a prototype of the central developmental phenomenon of localization. In the course of a day the essentially apolar zygote becomes differentiated into two grossly different regions and then cells: the tip-growing rhizoid cell, which serves for attachment, and the thallus cell, which will ultimately generate most of the plant (Fig. 3). The chief advantage of this system is the practically unpatterned state of the zygote; this state is shown not only by the appearance of the zygote and by the fact that a very wide variety of imposed vectors can determine the locus of rhizoid formation (L. F. Jaffe, 1968, Table I), but most convincingly of all by the fact that polarized light may cause up to half of the embryos to form rhizoids at two opposite loci (Fig. 4).

However, localization occurs normally in the absence of any imposed vector. Perhaps in this case the locus of sperm entry provides a necessary trace to localize rhizoid initiation [in the related form, *Cystoseira*, it has been shown to serve this function (Knapp, 1931)]; more likely no such trace is necessary: I *suppose* that if the egg could be activated by a uniform stimulus, and then cultured in a sufficiently vector-free environment, that some unavoidable molecular inhomogeneity would suffice to seed rhizoid initiation. In any case, since the effective vectors are so varied, and since no one of them is necessary, it would appear to be the best strategy to focus on the amplification process, which can attain the same embryonic pattern with almost any initiation trace, rather than on these vectors directly.

However, before doing that, I would like to point out one empirical rule which so far seems to describe the action of all such polarizing gradients: Suppose that a certain size gradient is imposed upon each of the cells in a large population. Then it turns out that the

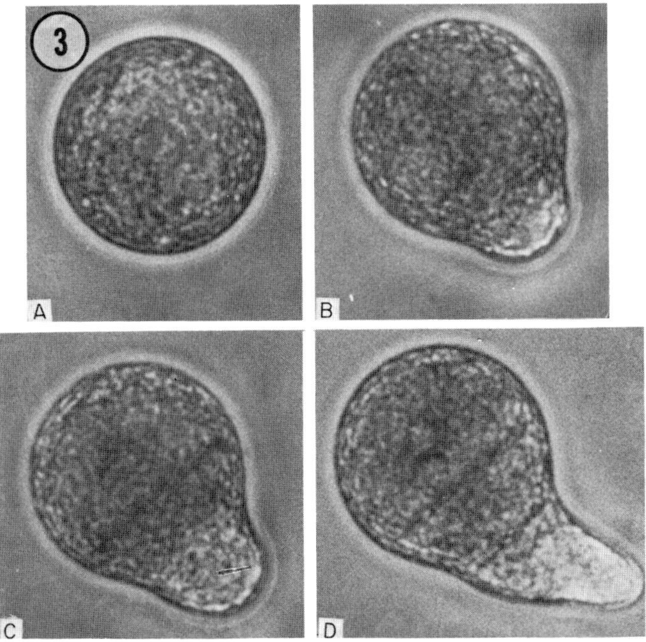

FIG. 3. Early development of *Fucus vesiculosus* (A) 4 hours after fertilization; (B) 16 hours—a rhizoidal bulge but no cell wall has formed; (C) 18 hours—a wall between rhizoid and thallus cell has formed; (D) 26 hours—further elongation and cell division. (Courtesy of Dr. B. Bouck.) From L. F. Jaffe (1968).

percent resultant polarization approximates the percent imposed gradient. This rule is well illustrated by the polarization of *Fucus* eggs by imposed hydrogen ion gradients (Bentrup *et al.*, 1967). It is also supported by measurements of the orientation by light of *Botrytis* and *Osmunda* spores (Jaffe and Etzold, 1962), the orientation by a flow-established gradient of a secreted growth stimulator in *Botrytis* spores (Müller and Jaffe, 1965), and *Fucus* eggs (Bentrup and Jaffe, 1968), as well as the orientation by a neighbor established gradient of a secreted growth inhibitor ("antirhizin") in *Fucus* eggs (Jaffe and Neuscheler, 1969). Although this rule is still unintelligible, it is nevertheless quite useful in analyzing the action of various vectors, and should ultimately prove to be a significant clue in understanding the early stages of the amplification process.

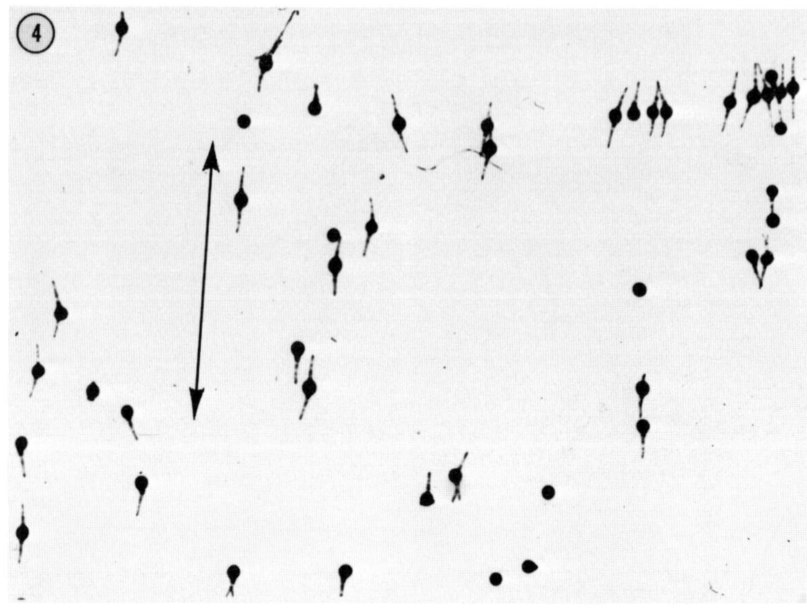

FIG. 4. Four-day-old embryos of *Fucus furcatus* cultured in plane-polarized light coming from both above and below. The arrow shows the direction of vibration of the electric vector. From Jaffe (1956).

THE AMPLIFICATION PROCESS IN FUCALES EGGS

Transcellular Electric Current as a Cause and Consequence of Localization

Turning now to the analysis of this amplification process, there are three broad questions to be asked:

1. What change in the whole system *initiates* the amplification process, that is, the process which converts the traces left by the environmental vectors first into an irreversibly if invisibly polarized, and then into a grossly differentiated, embryo?

2. How does the rhizoid anlage act back on itself so as to *augment* its own differentiation?

3. How does it act to *inhibit* other parts of the cell from going this same route, or using conventional developmental parlance, how does it dominate the rest of the system?

My best lead toward a solution to these questions was the finding a few years ago that as the amplification process begins, the embryo drives a substantial electrical current through itself (Jaffe, 1966).[1] Since it is only 100 μ long, probe resistances were too high to permit a direct current measurement; so currents had to be inferred from measurements of voltages and resistances. Furthermore, since it would only develop in media comparable to seawater and under relatively unconstrained conditions, the resistances could not be raised above about 10 kohms. Thus although the current densities proved to be high, of the order of 10 $\mu A/cm^2$, the available voltages were only fractions of a microvolt. This difficult measurement was attained only by putting several hundred developing eggs in series in a capillary, as shown in Fig. 5.

Figure 6 shows the results for *Pelvetia*, a close relative of *Fucus* with larger eggs, which can also be more reliably obtained. Several hours before the first eggs in the tube begin to germinate, i.e., before they initiate an outgrowth, a voltage appears across the tube which indicates the establishment of current loops through some or all of the eggs. No such voltage appears across control tubes in which the eggs are randomly oriented rather than parallel. The sign of the voltage indicates that current, considered as a flow of positive ions, enters each embryo at its growing or rhizoidal pole, traverses the cytoplasm, leaves the quiescent or thallus end, and returns through the medium to complete each loop. As long as normal development can be continued within the tube, which is until the 2- to 4-cell stage, the tube voltages and hence the inferred current intensities increase.

These currents are obviously a consequence of localization. Are they also a cause? I am inclined to believe that they are, for a number of reasons:

First of all the current and the process of axis fixation or determination seem to develop concurrently. Figure 7 shows the relationship of the time of photopolarizability to that of germination in *Pelvetia* eggs. The fall in photopolarizability is taken to indicate fixation of the axes of the eggs. The best reference time for comparing this

[1] Another promising lead to the localization problem in the Fucales eggs, recently discovered, is evidence for extracellular chemical controls. This lead comes from the so-called group effects. The Fucales eggs usually tend to germinate toward each other, but under certain conditions away from each other. The evidence now indicates that these interactions are mediated by diffusion of (at least) two substances, a rhizoid stimulator, rhizin, and an inhibitor, antirhizin (L. F. Jaffe, 1968). However, time and space forbid us to pursue this line on this occasion.

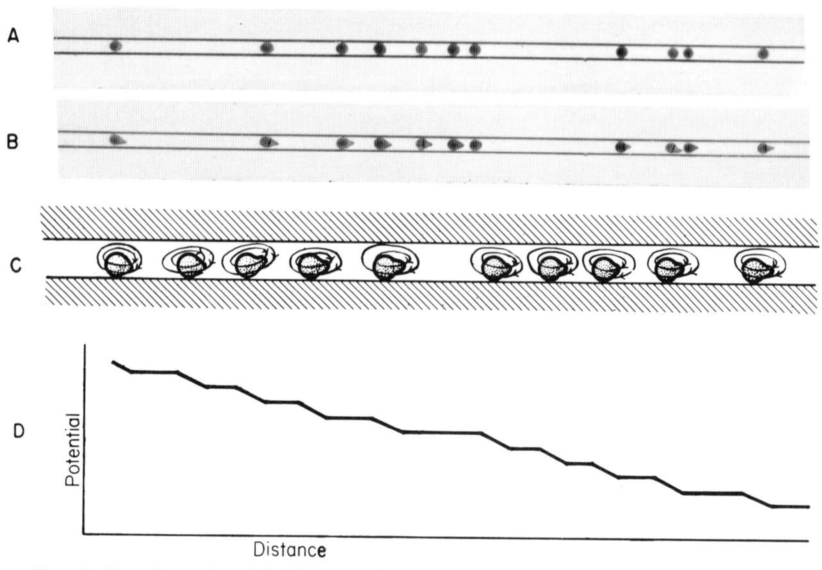

FIG. 5. Eggs in series. (A) Photograph of eggs in part of a 100 μ bore capillary before germination. Eggs are 75 μ in diameter. (B) Same eggs 26 hours after fertilization. (C) Schematic view of inferred current pattern in a tube. (D) Schematic graph of inferred change of potential along the tube. From L. F. Jaffe (1968).

curve with the tube voltage curve is probably the time that germination begins, or more precisely what I call the rise time (defined as the intersection of the linear part of the sigmoid germination curve with the abcissa). At this time, the tube voltage, and hence the average egg current, had risen to about a third of the plateau value reached when all the eggs have germinated but none has yet divided. On the other hand, photopolarizability has not fallen by a third until some hours *after* the rise time. The photopolarizability and current curves were necessarily obtained under somewhat different conditions. Moreover, they were done on different batches of eggs. Considering this as well as the subtle pitfalls of inferences from inhomogeneous populations, considerable caution must be used in comparing these curves. Nevertheless I believe it to be reasonably conservative to conclude for now that the development of current and of irreversibility are essentially concurrent phenomena.

Secondly, these eggs *can* be polarized by imposed fields or potassium ion gradients of as little as 15 mV or 15 m*M* per egg diameter, respectively, the rhizoids tending to form toward the higher poten-

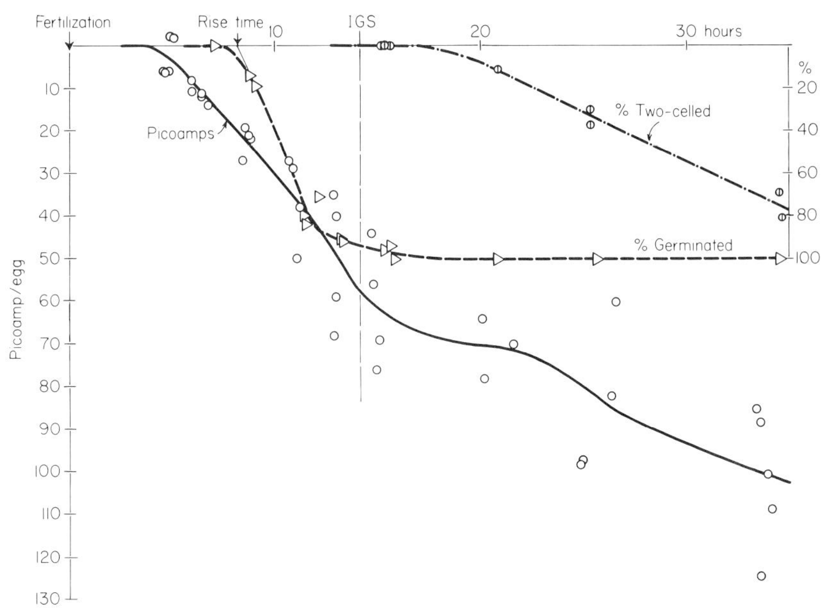

FIG. 6. Time course of currents flowing through developing eggs of *Pelvetia fastigiata*. From L. F. Jaffe (1968).

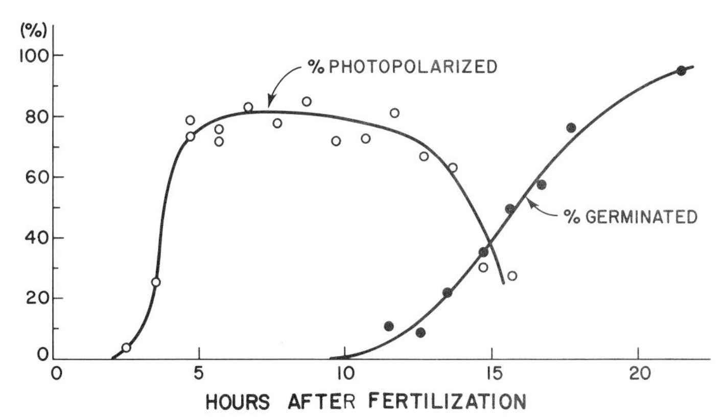

FIG. 7. Relationship of the time of photopolarizability to that of germination of *Pelvetia fastigiata* eggs. From L. F. Jaffe (1968).

tial or higher (K) (Lund, 1923; Bentrup *et al.*, 1967). Figure 8, taken from Lund's pioneering study, illustrates the point. These external agents drive currents through the egg in the same direction as the developing egg-driven current, and are probably of comparable size to it (L. F. Jaffe, 1968, p. 318); therefore their ability to polarize supports a causal role for the egg-driven current. Furthermore, while comparable fields can *not* reverse the polarity of eggs which have passed the stage of sensitivity to other vectors such as unilateral light, they *do* shift the growth point of already formed rhizoids towards the positive pole (Bentrup, 1968).

Bentrup has observed that, under certain conditions, imposed fields induce rhizoids to form toward the negative pole. He therefore challenges my interpretation of the egg-driven currents, arguing that they are only by-products of a localized membrane change and do not act to polarize the egg's interior (Bentrup, 1969a). However, such countereffects are only induced by fields imposed long before the egg-driven currents begin or irreversibility develops; hence they offer no evidence that localization can occur while current flows against the natural direction. Probably these early fields act by distorting the *extracellular* rhizoid-inhibiting gel; this can be inferred

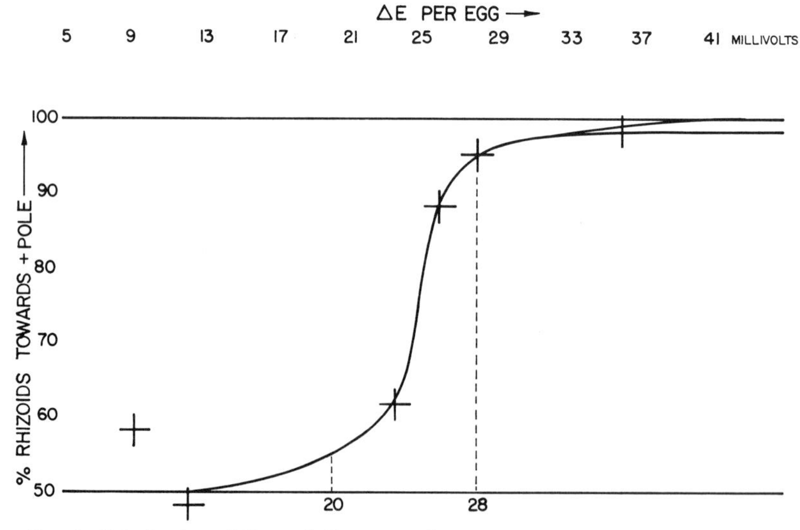

FIG. 8. Polarization of *Fucus inflatus* eggs by imposed electrical fields. From Lund (1923).

from the ability of rapid flow, applied at this early stage, to induce later rhizoid initiation upstream (Bentrup and Jaffe, 1968).

The third, and in my mind most important, reason for viewing the current as a cause as well as a consequence of localization lies in a consideration of just what the possible consequences of a transcellular current of such a magnitude could be. However, before doing that, it seems best first to consider the mechanism that may drive it.

The Transcellular Current and the Egg's Membrane Potential

General cell physiology tells us that the current must be driven by batteries in the cell membrane, specifically by a relative depolarization at the rhizoid pole. Since the only visible change during rhizoid initiation occurs there, the obvious guess is that the primary membrane change occurs there, too, and indeed that the current is caused by an absolute depolarization localized at the rhizoidal pole.

Consideration of fluctuations in the tube voltage suggests that this depolarization is not only localized but episodic and, moreover, quite large (L. F. Jaffe, 1968).

Cell physiology tells us that episodic depolarization usually arises from a change in the membrane from a so-called resting state, whose potential is controlled by its high potassium conductance, to an active state whose potential is controlled by its high conductance to some other ion. This is true whether the non-potassium ion is sodium as in squid axon (Hodgkin, 1958), calcium as in the barnacle muscle as well as *Paramecium* (Hagiwara and Naka, 1964; Naitoh and Ecker, 1969), or chloride as in *Nitella* (Mullins, 1962).

With these considerations in mind, how may the current's development be reflected in the egg's membrane potential? Consider Fig. 9.

Suppose that in some, probably small, region of the cell, namely that of the rhizoid anlage, the membrane's electromotive force falls by ΔE. It is easily shown, then, that the consequent fall in membrane potential, ΔV, will be given by:

$$\Delta V = \Delta E[R/(R' + R)] \tag{1}$$

where R is the resistance of the membrane region which remains in the resting state, and R' that of the activated part. Unless R' were very much larger than R, a substantial change in the membrane potential would be expected to accompany current initiation.

Figure 10 shows a representative recent measurement of the

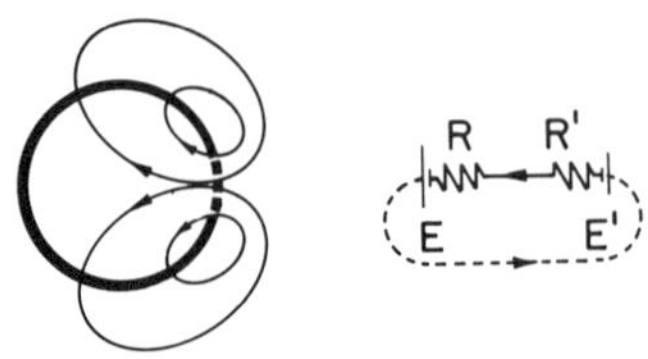

FIG. 9. The effect of a more or less depolarized patch on the eggs' average membrane potential. The electromotive force of the large unchanged region is E; that of the patch falls by ΔE to E'. Current will flow as shown, and the membrane potential will fall by $\Delta E \times [R/(R + R')]$.

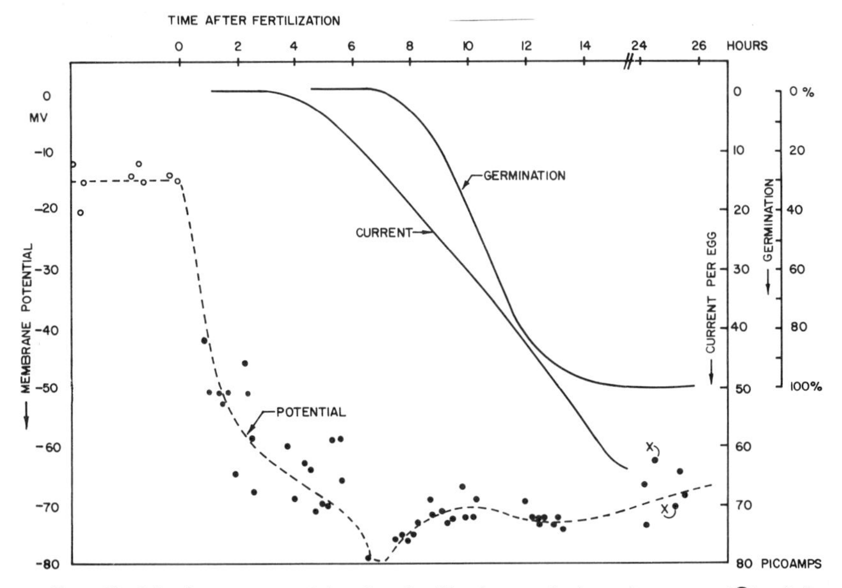

FIG. 10. Membrane potentials of unfertilized eggs (○), and zygotes (●) of the fucaceous alga *Pelvetia fastigiata* in artificial seawater at 15°C. In two cases the potentials of the rhizoid (×) and thallus cell (●) of the same embryo were measured separately. The figure shows a typical run of a synchronously fertilized batch. The germination rate and the transcellular current are replotted from L. F. Jaffe (1968) for comparison.

developing *Pelvetia* egg's membrane potential made by Dr. Manfred Weisenseel in my laboratory (Bentrup, 1969b, has comparable data). We consistently find the membrane potential to reach a peak of about 80 millivolts (inside negative) at about seven hours after fertilization. The course of current development as well as the germination of these eggs (from L. F. Jaffe, 1968) is put on the same graph for comparison. Considering that the current represents an average current per egg in an asynchronously developing population, we would guess that the peak potential *precedes* the start of the current through each egg. It certainly precedes its germination.

In any case, we consistently find the illustrated fall of about 5–10 mV, at about the time that the current starts. This important datum supports the inference of localized depolarization.

Table 1 shows some preliminary measurements, again made by Weisenseel, of the immediate responses of the egg's membrane potential to sudden changes in the ionic composition of the medium. The observed responses at 7–8 hours indicate that the peak voltage is close to a pure potassium potential. The responses to potassium changes are close to those predicted by the Nernst equation (at 15°C, a tenfold change corresponds to 58 mV). There is no detectable response to large changes in Na^+, Cl^-, or Ca^{2+}. Moreover, we have made flame photometric measurement of the egg's composition at this stage; they indicate concentrations of potassium and sodium of 140 and 15 mmoles per liter of cytoplasm, respectively. Considering that much of the cytoplasm is undoubtedly nonaqueous, these data are likewise consistent with a membrane potential entirely governed by its potassium conductance.

After this peak, however, there is an interesting change in this immediate potassium response. The response to a tenfold increase remains at about 50 mV. However, there is a 5- to 10-mV reduction in the response to a tenfold decrease.

What does that datum mean? I suspect this: that a region of mem-

TABLE 1

IMMEDIATE RESPONSES OF MEMBRANE POTENTIAL OF *Pelvetia fastigiata* EGGS TO SUDDEN CHANGES IN EXTERNAL IONIC CONCENTRATIONS[a]

Ion	Change	Activated			Polarized		
		Stage (hr): Unfert.	1–2	4–6	7–8	11–13	25
K^+	10 ×		−21	−41	−47	−48	−47
	1/10 ×	0	+15	+23	+40	+33	+34
Na^+	1/10 ×	+20	+25	0	0	0	0
Ca^{2+}	1/10 ×		−5	−7	0	−2	−3
Cl^-	1/10 ×	−5	−9	0?	0	0?	0?

[a] A plus sign indicates hyperpolarization (in millivolts). T = 15°C.

brane develops at the rhizoid pole the state of which is potential sensitive. Above a certain critical potential, this critical potential being a bit below the average membrane potential at that state—*above* this critical potential, this membrane region shifts to its activated, nonpotassium, and low electromotive force state.

This in turn plainly suggests an initial answer to our first question. What triggers the localization process? Perhaps it is a conjunction of two changes: First the observed increase in membrane potential, and second perhaps the development of a sensitivity to this potential; at the trigger point, the potential exceeds the critical potential, and the amplification process begins.

Before suggesting answers to our other two questions about localization, let us look at an interesting by-product of this study, namely the membrane changes that result from sperm activation.

The membrane of the unfertilized egg is radically different from that of the developing zygote. Its potential is very low, only 15 mV as compared to the 60–80 mV found in the embryo. Its potential shows no immediate potassium dependence, but a considerable response to sodium and chloride changes. These latter continue until about an hour after fertilization. While this analysis is quite incomplete, it is already clear that the state of the membrane at this stage bears some resemblance to that of the unfertilized and recently fertilized frog egg (Morrill *et al.*, 1966).

Transcellular Current and Self-electrophoresis

Now I wish to speculate a bit as to *how* the transcellular current may be a cause as well as a consequence of localization.

Three ways seem plausible:

1. The popped balloon hypothesis. This is a simple and attractive mechanism to explain how one rhizoidal region inhibits others from starting. The notion is that just as one hole in a balloon prevents others by lowering the air pressure, so one rhizoidal region blocks others by keeping the membrane potential down. Such electrical communication would be essentially instantaneous.

2. Ion gradient hypothesis. The above considerations of the mechanism of current propulsion strongly suggest that some cation other than potassium enters the growth pole while potassium leaves at the antipode. If this were true, and if a sufficient transembryonic diffusion barrier to the "other cation" exists, e.g., binding to an immobile cell constituent, then accumulation of the other cation could amplify the changes at the rhizoid pole through a variety of chemical mechanisms. The tube voltages implied that at least 100 picoamperes enters the rhizoidal region in the period around the first cell division. One hundred picoamps entering a region 30 μ × 30 μ × 10 μ deep would turn over and could displace all the small cations there in less than an hour. Thus this could be a fairly rapid mechanism for answering our second question: How does rhizoid differentiation snowball?

3. *Self-electrophoresis*. In traversing the cytoplasm the current will generate a field that may significantly localize negatively charged molecules or particles toward the growth point (or, if there are any, positively charged ones toward its antipode).

In my first analysis, I estimated this transcytoplasmic field to be about 1 to 10 mV/cm. This was simply done on the basis of the current density estimated from the tube voltages, a high frequency measurement of the intracellular resistivity (it proved to be about 200 ohm cm), and Ohm's law (Jaffe, 1966, 1968). These are small fields; nevertheless, hours are available for this egg's development and the persistence of stratification in the centrifuged egg indicates that mixing through streaming does not occur (Whitaker, 1940; Lowrance and Whitaker, 1940). However, an inevitable limitation upon segregation by such small fields is the leveling action of diffusion. Analysis indicates that at the equilibrium between electrophoresis and back diffusion, the gradient, G, of some component is given by the field strength, $\mathcal{E}$, multiplied by the ratio of electrophoretic mobility to diffusion constant, i.e.

$$G = (m/D)\mathcal{E} \tag{2}$$

For fields of 10 mV/cm, this reaches 10% per egg diameter for particles of the order of 100 Å *or larger*. Provided that the frictional coefficient of the cytoplasm for some component did not greatly exceed that of water, such an equilibrium would be reached in an hour or less. One may conclude that self-electrophoresis driven by such "ohmic" fields is another plausible mechanism whereby the current might effect localization.

Our new data and further consideration now suggest that the current may establish much larger fields than 10 mV/cm through a non-ohmic mechanism. For it now seems clear that the cations flowing into the growth point are in substantial part ions *other* than potassium, perhaps calcium and sodium. It is well known that these other ions, particularly calcium, are far less mobile in cytoplasm than in water.

Thus direct observations of the mobility of calcium in squid axons showed it to have no more than 2% of its mobility in water and possibly much less (Hodgkin and Keynes, 1957).

Let us consider the electrical consequences of the entry of locally immobilized cations. It will suffice for this purpose to think of the cytoplasm as an anionic gel. Local binding, then, will set up a fixed charge gradient and thus a field. Such a field would pull mobile, negatively charged components toward the cation entry region of the membrane.

How large might it be? Both experiment and theory indicate that it may be remarkably large.

First consider some measurements that seem to demonstrate the presence of very large potential differences originating in fixed charge gradients within various living cells. In a recent paper, Bruce and Christiansen (1965) report the reversible electrical effects of the anesthetics halothane and ether upon the giant amoeba, *Chaos chaos*. These agents somehow cause its cytoplasm to segregate, rapidly and reversibly, into a central, granular phase and a peripheral, clear one. Electron microscopy showed no membrane between these phases. Nevertheless the peripheral phase proved to be 65 *millivolts* positive with respect to the central one. While its exact ionic basis is quite unknown, there seems to be little doubt that this large potential difference is due to a fixed charge gradient. Moreover, comparably large potential differences have been reported between the inner and outer regions of the cytoplasm of both fish eggs and frog eggs under normal conditions (Hori, 1958; Morrill *et al.*, 1966). Since currents cannot move exclusively into the cell interior, and an ohmic field is thus excluded, and since no continuous membrane seems to have lain between the points proved in these experiments, the large radial potential gradients in these eggs seem likewise to be explicable only by way of fixed charged gradients of some sort.

Now let us consider the fields that may be set up by the influx of locally bound cations into an anionic gel on a theoretical basis. A so-called Donnan type potential will be set up (Davies and Rideal, 1963, p. 80). By repeating the derivation of the Donnan potential, with simplifications appropriate to this situation, one can show that the potential difference, E, between two points in an anionic gel is given by:

$$E \simeq 0.25\,\bar{p} \text{ (millivolts)} \tag{3}$$

where $\bar{p}$ is the *percent* difference in fixed charge concentration between these points *provided* only that $\bar{p}$ is small and that most of the anions are fixed (see Appendix for the derivation.)

Using this equation, let us consider the consequence of a current density, δ, of locally bound cations, entering a region of membrane for a time, t, and penetrating a distance, $\triangle$ before being bound. Let us take the fixed anionic charge density to be about 0.2 M, making the rough assumption that it is given by the combined potassium and sodium measured in the *Fucus* egg. Then one can easily show that

$$\bar{p} \simeq 5 \ \delta t/\triangle \tag{4}$$

$$E \simeq \delta t/\triangle \quad \text{(millivolts)} \tag{5}$$

$$\mathcal{E} \simeq E/\triangle = \delta t/\triangle^2 \ \text{(millivolts/cm)} \tag{6}$$

The current density entering the developing *Fucus* egg is estimated to be about 2 μA/cm^2. If all these ions were locally bound, how long would they take to set up a field of 1 V/cm? For a penetration depth $\triangle$ of 3 μ, this is less than 1 minute.

Evidently then the development of large fixed-charge fields within the cytoplasm near the growth point is a very attractive mechanism for amplifying a change at this pole. In particular, this would seem to be an attractive mechanism both for accumulating wall precursor vesicles near the growth point and for effecting their secretion.

HOW TO TEST THE ELECTRICAL HYPOTHESES

I believe some new techniques must be developed to effectively test these hypotheses. Two such are under development:

1. Single-cell ammeter. The multicell or series technique for measuring currents through eggs has severe limitations. Among these are a very low temporal resolution due to developmental asynchrony, absolutely minimal spatial resolution, and the difficulties of applying the method to eggs which are not photopolarizable and self-sticking. Accordingly, I am developing a new method, based upon a concept of Davies (1966) which should allow the detailed mapping of the current pattern through any egg developing in an aqueous environment.

The essential idea is to vibrate a capacitatively coupled probe normal to each investigated element of the egg's surface and lead the output to an amplifier locked into the vibration frequency. The amplifier's output should indicate the voltage gradient normal to the surface element; this together with the resistivity will yield the current flux through that element.

This system should provide radical improvement over the multiegg one. For, I can now report that we can make probes of 20 μ diameter which, when stationary in seawater, yield an r.m.s. noise value of only *1 nanovolt* (nV) at 100 cycles per second when averaged over 30 seconds. My calculations suggest that little or no unavoidable additional noise will be introduced by vibration at this frequency. I thus expect to produce an instrument capable in seawater of nearly 1 nV voltage resolution together with 20 μ spatial resolution, and 30 second temporal resolution.

One nanovolt resolution corresponds to about 20 nA/cm^2 in seawater. A rough comparison of the resolving power found in the multiegg system and expected in the new, vibrating probe system is given in the accompanying tabulation.

	Δt	Δl (μ)	$\Delta \delta$ (μA/cm^2)
Multiegg system	2 hours	100	2
Proposed vibrating probe	30 sec	20	0.02

2. *Cylindrical track autography*. At least one reliable technique for intracellular ion localization does exist (Winegrad, 1968), but it involves a very lengthy and laborious combination of freeze drying, vapor fixation, thin sectioning, and radioautography. Hence we have started to develop a radically shorter and simpler technique (Robinson and Jaffe, 1969). The essential idea is to observe the tracks produced in, and sufficiently normal to, a cylindrically shaped emulsion (Bonetti and Occhialini, 1951) surrounding a capillary containing whole, deep frozen eggs labeled with appropriate high energy β-emitters such as ^{42}K, ^{24}Na, ^{47}Ca, ^{36}Cl, and ^{28}Mg. Calculations indicate that the level of track origin will be resolvable to within a few microns. Moreover, it should be possible to complete an experiment within a few days.

DIRECT MEASUREMENTS OF LONGITUDINAL CYTOPLASMIC FIELDS

A literature search yields three isolated, but apparently reliable reports of such fields (Table 2). All these measurements were made directly with salt-filled intracellular micropipettes; in all cases the

TABLE 2
DIRECT MEASUREMENTS OF LONGITUDINAL CYTOPLASMIC FIELDS

Cell	Electropositive end of cytoplasm	Gradient (V/cm)	Reference
Fish melanophore	Granule receiving	1	Kinosita (1953, 1963)
Fish iridophore	Granule receiving	1	Kinosita (1953, 1963)
Neurospora hypha	Growing	1	Slayman and Slayman (1962)
Amoeba proteus	Advancing	1	Bingley (1966)

end toward which the grosser components of the cytoplasm moved was electropositive; in all cases the measured fields were of the order of 1 V/cm. An older study of Watanabe *et al.* (1937) seems to have given about the same result, although with less reliable techniques, in *Didymium* plasmodia.

These directly measured fields are in the same direction as that inferred for the *Fucus* embryo. They are certainly large enough to segregate cell constituents. These reports thus offer substantial support for self-electrophoresis as an intracellular segregation mechanism. Indeed, Kinosita (1953) proposed exactly such a mechanism for moving pigment granules in melanophores and iridophores.

TIP GROWTH AS LOCALIZED SECRETION

The origin and extension of the rhizoid in the Fucales egg can be viewed as a case of tip growth. It has become clear in recent years that plant cell wall growth and tip growth in particular is in large part effected by the secretion of wall materials in minute vesicles, a process sometimes called reverse pinocytosis (Pickett-Heaps and Northcote, 1966). This consideration focuses our attention upon vesicle secretion more generally. Evidence is rapidly accumulating that such secretion, whether it be of neurotransmitters, hormones, enzymes, or even of fertilization membrane precursors is peculiarly, immediately, and locally dependent upon calcium (Hales and Milner, 1968; Katz and Miledi, 1965; Castañeda and Tyler, 1968).

This recognition makes one wonder whether calcium plays some closely related role in wall growth. The positive chemotropism of several species of pollen toward calcium (Rosen, 1964) may illustrate such a role. Any direct action of calcium on the wall would be expected to increase the wall's rigidity and thus result in negative chemotropism. Thus, in causing positive chemotropism in pollen tubes it may be working in a different manner, e.g., by locally speeding vesicle secretion; through some as yet unknown but perhaps very general mechanism.

These thoughts then suggest that calcium may be the "other ion" or at least one of the "other ions" that apparently enter the growing rhizoid tip of the developing *Fucus* egg.

On the other hand, they suggest a general mechanism for the action of calcium on secretion.

May it not be that calcium traverses the membrane, is then fixed just inside of it, and thus generates a fixed charge field that pulls the secretion vesicles to the plasma membrane? One surprising finding that seems to be explained by this mechanism is the observation of Miledi and Slater (1966) that calcium injected into a neuronal synaptic ending in the squid fails to elicit transmitter release, even though it does so if applied extracellularly. This would be expected on the field theory because injected calcium would be initially more concentrated at some point well within the cytoplasm than immediately below the membrane. Unless some special ad hoc assumption were made the diffusion of such injected calcium might soon make it *as* concentrated just beneath the membrane as in a deeper position, but would not reverse the gradient. Thus, if anything, injected calcium would produce a field that would pull vesicles away from the cell membrane, not toward it. A test of this field theory might be to measure the temperature dependence of the synaptic delay; if it were sufficiently low, it would argue for such a physical mechanism.

CONCLUDING REMARKS

It is known that the two faces of polarized epithelia in the developed organism generally have radically different membranes. In frog skin, for example, the inner face is primarily potassium conductive; the outer, sodium conductive (Üssing, 1965). And now it turns out that even *Paramecium* has a comparable membrane polarity: The rear pole is potassium conductive; the front, upon stimulation at least, can become calcium conductive (Naitoh and Ecker, 1969).

These membrane gradients have been viewed primarily in terms of their function in acting upon the environment. Here it is proposed that they also act to localize the inner constituents of cells, and self-electrophoresis is the mechanism emphasized.

SUMMARY

1. Gene-centered views of development are challenged. It is argued, rather, that development proceeds centripetally: the surface changes first and the nucleus last.

2. Studies of localization in the developing *Fucus* egg are reviewed. It is tentatively concluded that the primary and controlling change is one from potassium to nonpotassium conductance at the rhizoid pole.

3. It is suggested that one important means whereby such a localized membrane change can organize inner components is through self-electrophoresis. In particular, the possible role of fields generated by the entry of locally bound cations such as calcium is emphasized.

APPENDIX I: DONNAN POTENTIAL BETWEEN TWO SIMILAR FIXED-CHARGE PHASES BEARING FEW MOBILE IONS OF THE FIXED CHARGES' SIGN

Consider two fixed charge-bearing phases, 1 and 2. Each contains a fixed charge concentration, X^-; a concentration of mobile cations, K^+ and of mobile anions, Cl^-. X_2^- slightly exceeds X_1^-. $p = \Delta X/X$. Cl^- is small compared to X^-. The equilibrium potential difference is E. (For simplicity, the indicator of each component's charge will be left off the symbols in the following derivation.

Electrical neutrality requires that:

$$X_1 + Cl_1 = K_1 \tag{1}$$

$$X_2 + Cl_2 = K_2 \tag{2}$$

The Boltzmann condition requires that:

$$K_2/K_1 = Cl_1/Cl_2 = e^{E/RT} \tag{3}$$

$$\therefore K_2 = K_1 Cl_1/Cl_2$$

Put (1) in (3):

$$K_2 = Cl_1\,(X_1 + Cl_1)/Cl_2 \tag{4}$$

Combine (2) and (4):

$$X_2 + Cl_2 = Cl_1\,(X_1 + Cl_1)/Cl_2$$

Solve for Cl_2:

$$Cl_2 = \tfrac{1}{2}(-X_2 + \sqrt{X_2^2 + 4Cl_1\,(Cl_1 + X_1)}$$

For small Cl_1:

$$Cl_2 = Cl_1\,X_1/X_2 \tag{5}$$

$$\therefore Cl_1/Cl_2 = X_2/X_1$$

Substitute (3): $$E = RT \ln X_2/X_1 \simeq RTp$$

$$\therefore E \simeq 25\, p \text{ (millivolts)} \tag{6}$$

ACKNOWLEDGMENTS

The investigations in the author's laboratory which are described here are being supported by NSF grant GB7048.

The critical instruction in electrophysiological techniques provided by Dr. William Pak and Dr. Moto Murikami is gratefully acknowledged; so, too, are valuable discussions with Dr. M. V. L. Bennett.

REFERENCES

ADLER, J. (1969). Chemoreceptors in bacteria. *Science* in press.

BAUMANN, G. (1968). Zur Wirkung des Juvenilhormons: Elektrophysiologische Messungen an der Zellmembran der Speicheldrüse von *Galleria mellonella*. *J. Insect Physiol.* **14,** 1959–1976.

BEISSON, J., and SONNEBORN, T. M. (1965). Cytoplasmic inheritance of the organization of the cell cortex in *Paramecium aurelia*. *Proc. Natl. Acad. Sci. U.S.* **53,** 275–282.

BENTRUP, F. W. (1968). Die Morphogenese pflanzlicher Zellen im elektrischen Feld. *Z. Pflanzenphysiol.* **59,** 309–339.

BENTRUP, F. W. (1969a). Zur Funktion der Zellmembran bei der Cytomorphogenese. *Ber. Deut. Botan. Ges.* **7,** 311–314.

BENTRUP, F. W. (1969b). Membrane potential measurements on fertilized *Fucus* eggs. *Naturwissenschaften* **56,** 331–332.

BENTRUP, F. W., and JAFFE, L. F. (1968). Analyzing the 'group effect': Rheotropic responses of developing *Fucus* eggs. *Protoplasma* **65,** 25–35.

BENTRUP, F. W., SANDAN, T., and JAFFE, L. F. (1967). Induction of polarity in *Fucus* eggs by potassium ion gradients. *Protoplasma* **64,** 254–266.

BINGLEY, M. S. (1966). Further investigations into membrane potentials in amoebae. *Exptl. Cell Res.* **43,** 1–12.

BONETTI, A., and OCCHIALINI, G. P. S. (1951). Cylindrical emulsions. *Nuovo Cimento* **8,** 725–727.

BRIGGS, W. R. (1963). The phototropic responses of higher plants. *Ann. Rev. Plant Physiol.* **14,** 311–352.

BRUCE, D., and CHRISTIANSEN, R. (1965). Morphologic changes in the giant amoeba *Chaos chaos* induced by halothane and ether. *Exptl. Cell Res.* **40,** 544–553.

CASTAÑEDA, M., and TYLER, A. (1968). Adenyl cyclase in plasma membrane preparations of sea urchin eggs and its increase in activity after fertilization. *Biochem. Biophys. Res. Commun.* **33,** 782–787.

CURTIS, A. S. G. (1963). The cell cortex. *Endeavour* **22,** 134–137.

DAVIES, J. T., and RIDEAL, E. K. (1963). "Interfacial Phenomena," 2nd ed. Academic Press, New York.

DAVIES, P. W. (1966). Membrane potential and resistance of perfused skeletal muscle fibers with control of membrane current. *Federation Proc.* **25,** 332.

DELBRÜCK, M. (1969). *Phycomyces:* Orientation of receptor pigment. *Bacteriol. Rev.* **33,** 134.

EDELMAN, I. S., BOGOROCH, R., and PORTER, G. (1963). On the Mechanism of action of aldosterone on sodium transport: The role of protein synthesis. *Proc. Natl. Acad. Sci. U.S.* **50,** 1169–1176.

ETZOLD, H. (1965). Der Polarotropismus und Phototropismus der Chloronemen von *Dryopteris Filix Mas* (L.) Schott. *Planta* **64,** 254–280.

FURSHPAN, E. J., and POTTER, D. D. (1968). Low-resistance junctions between cells in embryos and tissue culture. *Current Topics Develop. Biol.* **3,** 95–128.

GARCÍA-BELLIDO, A. (1966). Pattern reconstruction by dissociated imaginal disk cells of *Drosophila melanogaster*. *Develop. Biol.* **14,** 278–306.

GERISCH, G. (1968). Cell aggregation and differentiation in *Dictyostelium*. *Current Topics Develop. Biol.* **3,** 159–198.

HAGIWARA, S., and NAKA, K. (1964). The initiation of spike potentials in barnacle muscle fibers under low intracellular Ca . *J. Gen. Physiol.* **48,** 141–162.

HALES, C. N., and MILNER, R. D. G. (1968). The role of sodium and potassium in insulin secretion from rabbit pancreas. *J. Physiol.* (*London*) **194,** 725–743.

HAUPT, W. (1965). Perception of environmental stimuli orienting growth and movement in lower plants. *Ann. Rev. Plant Physiol.* **16,** 267–290.

HAUPT, W. (1966). Phototaxis in plants. *Intern. Rev. Cytol.* **19,** 267–299.

HAUPT, W. (1968). Die Orientierung der Phytochrom-Moleküle in der *Mougeotia*-Zelle. *Z. Pflanzenphysiol.* **58,** 331–346.

HENDRICKS, S. B., and BORTHWICK, H. A. (1967). The function of phytochrome regulation of plant growth. *Proc. Natl. Acad. Sci. U.S.* **58,** 2125–2130.

HERTEL, R., EVANS, M. L., LEOPOLD, A. C., and SELL, H. M. (1969). The specificity of the auxin transport system. *Planta* **85,** 238–249.

HODGKIN, A. L. (1958). Ionic movements and electrical activity in giant nerve fibres. *Proc. Roy. Soc.* (*London*) **B148,** 1–37.

HODGKIN, A. L., and KEYNES, R. D. (1957). Movements of labelled calcium in squid giant axons. *J. Physiol.* (*London*) **138,** 253–281.

HOLTFRETER, J. (1943). A study of the mechanics of gastrulation. *J. Exptl. Zool.* **94,** 261–318.

HORI, R. (1958). On the membrane potential of the unfertilized egg of the Medaka. *Embryologia* **4,** 79–91.

JACOB, F., and MONOD, J. (1963). Genetic repression, allosteric inhibition, and cellular differentiation. *In* "Cytodifferentiation and Macromolecular Synthesis" (*21st Symp. Soc. Study Develop. Growth*, M. Locke, ed.), pp. 30–64. Academic Press, New York.

JAFFE, L. F. (1956). Effect of polarized light on polarity of *Fucus*. *Science* **123,** 1081–1082.

JAFFE, L. F. (1960). The effect of polarized light on the growth of a transparent cell. *J. Gen. Physiol.* **43,** 897–911.

JAFFE, L. F. (1962). Evidence that the electric vector governs light absorption in vision, phototropism, and phototaxis. *Photochem. Photobiol.* **1,** 211–216.

JAFFE, L. F. (1966). Electrical currents through the developing *Fucus* egg. *Proc. Natl. Acad. Sci.* U.S. **56,** 1102–1109.

JAFFE, L. F. (1968). Localization in the developing *Fucus* egg and the general role of localizing currents. *Advan. Morphogenesis* **7,** 295–328.

JAFFE, L. F., and ETZOLD, H. (1962). Orientation and locus of tropic photoreceptor molecules in spores of *Botrytis* and *Osmunda*. *J. Cell Biol.* **13,** 13–31.

JAFFE, L. F., and ETZOLD, H. (1965). Tropic responses of *Funaria* spores to red light. *Biophys. J.* **5,** 715–742.

JAFFE, L. F., and NEUSCHELER, W. (1969). On the mutual polarization of nearby pairs of fucaceous eggs. *Develop. Biol.* **19,** 549–565.

JAFFE, M. J. (1968). Phytochrome-mediated bioelectric potentials in mung bean seedlings. *Science* **162,** 1016–1017.

JENKINSON, I. S. (1962). Bioelectric oscillations of bean roots. II. *Australian J. Biol. Sci.* **15,** 101–114.

JENKINSON, I. S., and SCOTT, B. I. H. (1961). Bioelectric oscillations of bean roots. I. *Australian J. Biol. Sci.* **14,** 231–237.

KATZ, B., and MILEDI, R. (1965). The effect of calcium on acetylcholine release from motor nerve terminals. *Proc. Roy. Soc.* (*London*) **B161,** 496–503.

KINOSITA, H. (1953). Studies on the mechanism of pigment migration within fish melanophores with special reference to their electric potentials. *Annotationes Zool. Japan:* **26,** 115–127.

KINOSITA, H. (1963). Electrophoretic theory of pigment migration within fish melanophores. *Ann. N.Y. Acad. Sci.* **100,** 992–1004.

KNAPP, E. (1931). Entwicklungsphysiologische Untersuchungen an Fucaceen-Eiern. *Planta* **14,** 731–751.

LOEWENSTEIN, W. R. (1968). Communication through cell junctions: Implications in growth control and differentiation. *In* "The Emergence of Order in Development" (27th *Symp. Soc.* for *Develop. Biol.*, M. Locke, ed.), *Develop. Biol.*, Suppl. 2, 151–183.

LOWRANCE, E. W., and WHITAKER, D. M. (1940). Determination of polarity in *Pelvetia* eggs by centrifuging. *Growth* **4,** 73–76.

LUND, E. J. (1923). Electrical control of organic polarity in the egg of *Fucus. Botan. Gaz.* **76,** 288–301.

MILEDI, R., and SLATER, C. R. (1966). The action of calcium on neuronal synapses in the squid. *J. Physiol.* (*London*) **184,** 473–498.

MONROY, A. (1965). "Chemistry and Physiology of Fertilization." Holt, New York.

MORRILL, G. A., ROSENTHAL, J., and WATSON, D. E. (1966). Membrane permeability changes in amphibian eggs at ovulation. *J. Cell Physiol.* **67,** 375–389.

MÜLLER, D., and JAFFE, L. F. (1965). A quantitative study of cellular rheotropism. *Biophys. J.* **5,** 317–335.

MULLINS, L. J. (1962). Efflux of chloride ions during the action potential of *Nitella. Nature* **196,** 986–987.

NAITOH, Y., and ECKER, R. (1969). Ionic mechanisms controlling behavioral responses of *Paramecium* to mechanical stimulation. *Science* **164,** 963–965.

PICKETT-HEAPS, J. D., and NORTHCOTE, D. H. (1966). Relationship of cellular organelles to the formation and development of the plant cell wall. *J. Exptl. Botany* **17,** 20–26.

RAVEN, C. P. (1967). The Distribution of special cytoplasmic differentiations of the egg during early cleavage in *Limnaea stagnalis. Develop. Biol.* **16,** 407–437.

ROBINSON, K. R., and JAFFE, L. F. (1969). A proposed new technique for autoradiography. *Third Intern. Biophys. Congr., Abstracts*, 305.

ROBISON, G. A. (1969). Cyclic AMP as a second messenger. *J. Reprod. Fertility* in press.

RODBELL, M., JONES, A. B., CHIAPPE DE CINGOLANI, G. E., and BIRNBAUMER, L. (1968). The actions of insulin and catabolic hormones on the plasma membrane of the fat cell. *Recent Progr. Hormone Res.* **24,** 215–254.

ROSEN, W. G. (1964). Chemotropism and fine structure of pollen tubes. *In* "Pollen Physiology and Fertilization" (H. F. Linskens, ed.), pp. 159–166. North-Holland Publ., Amsterdam.

SAXÉN, L., KOSKIMIES, O., LAHTI, A., MIETTINEN, H., RAPOLA, J., and WARTIOVAARA, J. (1968). Differentiation of kidney mesenchyme in an experimental model system. *Advan. Morphogenesis* **7,** 251–293.

SINYUKHIN, A. M., and BRITIKOV, E. A. (1967). Action potentials in the reproductive systems of plants. *Nature* **215,** 1278–1280.

SLAYMAN, C. L., and SLAYMAN, C. W. (1962). Measurement of membrane potentials in *Neurospora*. *Science* **136,** 876–877.

SMITH, L. D., and ECKER, R. E. (1969). Role of the oocyte nucleus in physiological maturation in *Rana pipiens*. *Develop. Biol.* **19,** 281–309.

STEINBERG, M. S. (1964). The problem of adhesive selectivity in cellular interactions. *In* "Cellular Membranes in Development" (*22nd Symp. Soc. for the Study Develop. Growth*, M. Locke, ed.), pp. 321–366. Academic Press, New York.

SUTHERLAND, E. W., OYE, I., and BUTCHER, R. W. (1965). The action of epinephrine and the role of the adenylcyclase system in hormone action. *Recent Progr. Hormone Res.* **21,** 623–646.

SWEENEY, B. M. (1941). Conditions effecting the acceleration of protoplasmic streaming by auxin. *Am. J. Botany* **28,** 700–702.

SZEGO, C. M., and DAVIS, J. S. (1967). Adenosine-3′,5′-monophosphate in rat uterus: acute elevation by estrogen. *Proc. Natl. Acad. Sci. U.S.* **58,** 1711.

TANADA, T. (1968). A rapid photoreversible response of barley root tips in the presence of 3-indoleacetic acid. *Proc. Natl. Acad. Sci. U.S.* **59,** 376–380.

TARTAR, V. (1962). Morphogenesis in *Stentor*. *Advan. Morphogenesis* **2,** 1–26.

ÜSSING, H. H. (1965). Transport of electrolytes and water across epithelia. *Harvey Lectures Ser.* **59,** 1–30.

WAGNÉ, C. (1965). The distribution of the light effect from irradiated to non-irradiated parts of grass leaves. *Physiol. Plantarum* **18,** 1001–1006.

WARD, P. A., COCHRANE, C. G., and MÜELLER-EBERHARD, H. J. (1965). The role of serum complement in chemotaxis of leucocytes *in vitro*. *J. Exptl. Med.* **122,** 327–346.

WATANABE, A., KODATI, M., and KINOSHITA, S. (1937). Beziehung zwischen der Protoplasmaströmung und den elektrischen Potentialveränderungen bei Myxomyceten. (In Japanese with a German summary.) *Botan Mag. (Tokyo)* **51,** 155–167.

WATERMAN, T. H., FERNANDEZ, H. R., and GOLDSMITH, T. H. (1969). Dichroism of photosensitive pigments in rhabdoms of the crayfish, *Orconectes*. *J. Gen. Physiol.* in press.

WHITAKER, D. M. (1940). The effects of ultracentrifuging and of pH on the development of *Fucus* eggs. *J. Cellular Comp. Physiol.* **15,** 173–187.

WINEGRAD, S. (1968). Intracellular calcium movements of frog skeletal muscle during recovery from tetanus. *J. Gen. Physiol.* **51,** 65–83.

DEVELOPMENTAL BIOLOGY SUPPLEMENT 3, 112–132 (1969)

Cellular Interaction in the Induction of Antibody Synthesis

FRANK L. ADLER AND MARVIN FISHMAN

Department of Immunology, The Public Health Research Institute of the City of New York, Inc., New York, New York

INTRODUCTION

The inductive phase of antibody formation embraces the time between injection of an antigen and the first appearance of measurable antibody. It may be as short as 7.5 minutes (Litt, 1967) or may extend to days or weeks (Nossal *et al.*, 1964), and is greatly affected by the nature and physical state of the antigen, the route by which it is administered, the species, strain, and the prior experiences of the experimental animal. Under appropriate conditions the antigenic stimulus may not result in overt antibody production but rather in the establishment of a state of sensitization (delayed sensitivity), or it may merely prepare the animal for an accelerated response to a second exposure to the antigen. Under still other circumstances, and again after an induction period (Weigle and Golub, 1967), a specific unresponsiveness (tolerance) to the test antigen may evolve. It is the purpose of this presentation to discuss some of the recent work which deals with cellular and molecular events that occur during this inductive period, with emphasis on the mounting evidence that complex cellular interactions may play a significant role. It is not our purpose to review the voluminous literature in detail; references cited have therefore been selected to illustrate specific points or to direct the reader to appropriate reviews. These include articles by Fishman (1969), Sulitzeanu (1968), M. Cohn (1967), Shands (1967), and Straus (1967).

THE ROLE OF PERITONEAL MACROPHAGES IN THE INITIATION OF A PRIMARY ANTIBODY RESPONSE *IN VITRO*

To facilitate studies on the inductive period of antibody formation, and in particular to examine the significance of cellular interaction, a system entirely *in vitro* offers numerous advantages. When we started our investigation it was clear from the experience of others that the addition of antigen to spleen, lymph node, or bone marrow cells from nonimmunized animals resulted in antibody formation only on rare and unpredictable occasions (Carrel and Ingebrigtsen, 1912).

Impressed by the evidence that cells of the reticuloendothelial system trap and store antigen for long periods (McMaster, 1961; Campbell and Garvey, 1963), and that blockade of this system interferes with antibody formation even though the cells in question are not important producers of antibody (McMaster, 1961; Jaffé, 1931), we set out to test the hypothesis (Ehrich, 1956) that phagocytosis of the antigen by appropriate cells may yield the effective immunogenic stimulus for the major source of antibody, namely cells of the lymphoid–plasma cell series. Antigens, usually bacteriophages, were therefore subjected to phagocytosis by peritoneal exudate cells rich in macrophages, and a cell-free extract prepared from such cells after homogenization was added to cultures of lymph node cells. Specific antibody production ensued in an appreciable number of experiments (Fishman, 1959, 1961), against a background of negative results if antigen alone or antigen mixed with an extract of nonstimulated macrophages was added to the cultures.

As a result of the findings by Campbell and Garvey (1963), which showed that antigen in highly immunogenic form was stored as a complex with ribonucleic acid (RNA), and because of our observation that streptomycin, which binds nucleic acids, and ribonuclease, which interfered in our system, we then substituted RNA extracts from antigenically stimulated macrophages for the crude material and found such preparations to be active in inducing antibody formation (Fishman and Adler, 1963; Fishman *et al.*, 1965a). Antibody appeared in the culture fluids in two waves, the first cresting on day 4 of the culture, the second on about day 12. Tests for sensitivity to mercaptoethanol, and the results of gradient centrifugation showed that the early antibody was IgM (19 S) and the late antibody was IgG (7 S). The sequential appearance of antibodies with similar specificities yet of different immunoglobulin classes has been frequently observed *in vivo*. It will be shown below that in our *in vitro* system each of the two classes is elicited by a distinct fraction of the macrophage RNA extract.

The immunogenic activity of RNA extracts from macrophages that have been incubated with antigen *in vitro* was promptly confirmed by others. Friedman *et al.* (1965) and Askonas and Rhodes (1965) called attention to the presence of antigen in such extracts, but they found that the biological activity was diminished as a result of RNase digestion. Bishop *et al.* (1967), Cohen and Raska (1967), Gottlieb *et al.* (1967), and Mosier and Cohen (1968) reported similar findings.

Recently it has been shown that cultures of teased spleen cells maintained under specified conditions can be induced to make hemolytic antibody against red blood cells in response to the addition of red cells to the culture (Dutton and Mishell, 1967; Marbrook, 1967). Such spleen cells can be separated into a population that adheres to glass, and into nonadherent cells. It has been shown that neither population alone will support antibody synthesis and that a primary interaction between antigen and the macrophage-rich adherent cells is essential for its initiation (Mosier, 1967).

THE RETICULOENDOTHELIAL SYSTEM AND ANTIBODY FORMATION *IN VIVO*

Extrapolation from *in vitro* experiments are open to the objection that conditions selected deliberately or attained by accident may be artifactual and thus irrelevant to antibody formation in the intact animal. It is therefore necessary to marshal the evidence which links the experiments just cited to what is believed to occur during the inductive phase of antibody formation in an animal that has been injected with antigen.

As already mentioned, blockade of the reticuloendothelial (RE) system, e.g., by the injection of carbon particles or other substances, has immunosuppressive effects, and cells of this system have been shown to store antigen for long periods of time. Among other circumstantial evidence that implicates the RE system as a supporting link in antibody formation is the phenomenon of peripolesis (Sharp and Burwell, 1960) in which lymphoid cells appear to adhere to a central macrophage; the finding of cytoplasmic bridges between such cells (Schoenberg *et al.*, 1964); and the immunosuppressive action of cytotoxic antimacrophage sera (Panijel and Cayeux, 1968). In addition, it has been found that high speed centrifugation of antigen solutions or their passage through an animal by injection can bring about separation of the protein solutions into aggregated or readily phagocytosed material which is highly immunogenic, and into a more soluble and less readily cleared fraction which, under proper conditions, engenders tolerance rather than antibody formation (Dresser and Mitchison, 1968; Frei *et al.*, 1965).

More directly in support of a key role for the macrophage in immunogenesis are findings which show that immunological maturation of the newborn animal can be hastened by the injection of compatible adult macrophages (Argyris, 1968), and that the unresponsiveness after total body X-irradiation which persists after lymphoid repopulation has occurred can be relieved by the administration of macrophages (Gershon and Feldman, 1968). There is also suggestive evidence that in some instances the genetically determined hyporesponsiveness to a given antigen may be traced to the macrophage. It has been possible to show that mice of inbred strains which make little or no antibody to the synthetic polypeptide GAT do respond to the injection of RNA extracted from macrophages of rabbits or "responder" mice that had been incubated with GAT (Pinchuck *et al.*, 1968).

Additionally, there are a number of studies (Mitchison, 1968; Unanue and Askonas, 1968) which show that some soluble protein antigens are more immunogenic after they have been incubated with macrophages than when they are injected as free antigen. Interestingly, this enhanced antigenicity is demonstrable in the course of primary immunization and not always in a secondary response to the test antigen.

It should be pointed out that cells of the reticuloendothelial system and in particular macrophages are not generally considered to be antibody producers (Nossal, 1966b). Nevertheless, there is much evidence that at least some of these cells can indeed mount an antibody response, possibly one of modest proportions (Harris *et al.*, 1966; Bussard, 1966; Noltenius and Chahin, 1969; see, however, Bendinelli, 1968).

Recent work by Garvey *et al.* (1967) shows that RNA–antigen complexes of high immunogenicity can be recovered, under the proper conditions, from blood and from urine of immunized animals. Their findings lend strength to similar observations of others, and it may be surmised that the cellular origin of some of this heterogeneous material is in the RE system. Worthy of attention are also the studies by Ford *et al.* (1966) which led these workers to conclude that ingestion of sheep red blood cells by macrophages is an essential step in the initiation of hemolysin production by lymphocytes. Ford *et al.* also noted that in some instances lymphocytes from specifically tolerant animals became producers of hemolytic antibody as a result of their exposure to the macrophages that had ingested antigen.

Also relevant is the exacting work of Nossal and his associates on antigen localization at sites of active antibody synthesis which has drawn attention to histiocytic cells that have variously been referred to as dendritic macrophages or reticular cells. It has been shown that such cells concentrate antigen on their processes where lymphoid cells come into intimate contact with the retained antigen. Whether this antigen concentration is causally related to antibody formation or is secondarily caused by it remains an unsettled question (Humphrey and Frank, 1967; Ada *et al.*, 1968).

RIBONUCLEIC ACID FROM ANTIGENICALLY STIMULATED MACROPHAGES AS THE IMMUNOGEN (VERTICAL CELL-TO-CELL TRANSMISSION)

We would now like to return to a discussion of the nature of the immunogenic RNA that can be extracted from macrophages after such cells have been exposed to antigen. In our *in vitro* system which, as already stated, yields antibody in two waves, it has been possible to show that IgM and IgG antibodies are elicited by two distinct fractions of RNA. The early 19 S antibody is formed in response to RNA which is synthesized in the macrophages after the addition of antigen (Fishman *et al.*, 1965b). Such RNA is free of detectable antigen, is highly sensitive to degradation by RNase, appears in response to minute doses of antigen, and can be separated from the other active fraction chromatographically on columns of methylated bovine serum albumin-Kieselguhr (MAK). Its molecular weight falls into the 8 S to 16 S class, and its biological activity is resistant to proteolytic enzymes, including pronase. The late 7 S antibody is induced by preformed macrophage RNA which is complexed to antigen or antigen fragments, and which has an apparent molecular weight corresponding to 23–28 S. This RNA is somewhat resistant to RNase, yet loses biological activity upon appropriate treatment with either RNase or pronase. It can be removed from the crude RNA by precipitation with specific antibody against the attached antigen, and it is formed in detectable amounts only after the addition to macrophages of amounts of antigen that are at least 1000-fold in excess over the amount required to incite formation of the RNA responsible for 19 S antibody formation (Fishman and Adler, 1967; Fishman *et al.*, 1968). The distinguishing properties of the two immunogenic fractions of macrophage RNA and some additional details are summarized in Table 1.

In work to be published we have been able to show that the RNA-antigen complex can be produced in a cell-free system consisting of solubilized antigen, RNA from nonstimulated macrophages, and cell sap. Our evidence suggests that the complex is formed with the aid of one or more enzymes and that electrostatic bonding, such as had been suggested (Roelants and Goodman, 1968) for other systems, is not involved. Antibody formed in cultures of lymph nodes in response

TABLE 1

CHARACTERISTICS OF THE IMMUNOGENIC RNA FRACTIONS RESPONSIBLE FOR EARLY (19 S) AND LATE (7 S) ANTIBODY FORMATION *in Vitro* AGAINST T2 BACTERIOPHAGE[a]

Characteristic	19 S	7 S
Presence of T2 antigen(s)		
Quenching of antigenicity by anti-T2	−	+
Precipitation of active RNA by anti-T2	−	+
Lability to pronase, 1 μg/100 μg RNA	−	+
5 μg/100 μg RNA	−	+
Dependence on antigen dose added to macrophages		
Ratio of T2:macrophage 1000:1	+	++
100:1	+	+
10:1	+	−
1:1	+	−
1:10	+	−
1:100	+	−
1:1000	+	−
Inactivation by ribonuclease		
1 μg enzyme:100 μg RNA	+	−
5 μg enzyme:100 μg RNA	+	+
Inhibition of formation after antigen addition		
Actinomycin D	+	−
Transfer of allotypic markers of macrophage donor	+	−
Estimated S value	8–10	23–28
Fractionation by MAK chromatography		
MAK I	−	−
MAK II	+	−
MAK III	−	+

[a] In the typical experiment, 10^9 macrophages from nonimmunized animals were incubated for 30 minutes at 37°C with 10^6 to 10^{12} plaque-forming units of T2 phage. RNA was extracted from the washed cells, and 100 μg of such RNA, untreated, treated as indicated in the table, or fractionated as shown, was added to cultures of lymph node fragments in 5 ml of Trowell's medium. Fluids were titrated for 19 S antibody on days 4–6 and for 7 S antibody on days 10–13 of culture, phage neutralization being used as an index of antibody formation. For other details see text.

to such RNA follows closely the kinetics of the 7 S response seen after addition of the purified RNA–antigen complex obtained from intact macrophages that had been incubated with antigen.

Gottlieb (1968) has isolated, by density gradient centrifugation, an RNA-antigen complex from macrophages that had reacted with antigen. This material was sensitive to pronase but rather resistant to

RNase. It appears to be similar to that fraction of our material which induces 7 S antibody formation in cultures of lymph node cells. Gottlieb has assayed the immunogenicity of his material in cultures of spleen cells, and for this and other reasons it has not yet been possible to see whether his and our materials are identical. While Gottlieb's active RNA fraction is thought to be preformed "antigen-carrier" RNA, other workers have observed the synthesis of new and apparently unique RNA in macrophages that have taken up antigen. Halac *et al.* (1965) described novel base ratios in RNA extracted from macrophages that had been incubated with bovine serum albumin, and Cohen and Raska (1967) found that mouse peritoneal exudate cells produced, upon antigenic stimulation, "new species" of RNA. Using a combination of radioactive labeling and DNA–RNA hybridization they arrived at the conclusion that, while both adherent (macrophage rich) and nonadherent cells of the peritoneal exudate made "new species" of RNA, only the latter cells made RNA which was both of a new species *and* antigen-specific. These authors tentatively ascribed messenger function to the two classes of freshly synthesized RNA but were not able to specify the function more precisely.

One important lead toward an understanding of the functions of the two RNA fractions in our work has come from experiments in which rabbits of different allotypes were used as the donors of macrophages and lymph node cells, respectively (Adler *et al.*, 1966). These experiments have shown that the 19 S antibody formed in our cultures bears the allotypic markers of the macrophage donor on both the light (L) and heavy (H) polypeptide chains of the molecules. Typical results of an experiment in which the transfer of L-chain markers was studied are shown in Table 2. This observation is most readily explained by postulating that the RNA in question is informational RNA which codes for both antigenic and antibody specificities. Since it has been shown that amino acid sequences which specify these two attributes are intimately associated (Koshland, 1967) and that the H and L chains are formed on separate polysomes (Askonas and Williamson, 1967; Lennox *et al.*, 1967), we are led to believe that our 19 S antibody-producing cells must be the recipients of two "messenger" RNA molecules, one each for the L and H chains or for the amino-terminal "variable" portions of these chains. Since approximately 100 μg of RNA is required per tissue culture plate, such requirement for a "double hit" could well be met under the

TABLE 2

ALLOTYPIC SPECIFICITY OF ANTIBODY AGAINST T2 PHAGE PRODUCED IN CULTURE[a, b]

Expt. No.	Donor of		Incubation (days)	Neutralization of T2 phage T2 fluids preincubated with:		
	Peritoneal exudate cells	Lymph node fragments		Normal rabbit serum (%)	Anti-4 (%)	Anti-5 (%)
1	b^4b^4	b^5b^5	5	53	10	40
			10	38	34	0
2	b^4b^4	b^5b^5	5	34	3	44
			12	27	36	17
3	b^5b^5	b^4b^4	5	60	50	19
			13	38	0	35
4	b^5b^5	b^4b^4	4	44	39	0
5	b^4b^4	b^5b^5	5	34	4	30

[a] Reproduced from Adler *et al.* (1966). We thank the publishers, The Williams & Wilkins Co., Baltimore, Maryland, for permission to use this table.

[b] The allotypic specificity of antibody found in tissue culture fluids after incubation for the number of days indicated was determined with the aid of typing sera specific for either the b^4 or the b^5 marker. Normal rabbit serum served as the control. The allotypic specificity of the phage-neutralizing antibody is apparent from the reduction of activity that results from preincubation with the appropriate typing serum. Experimental details are described in the original paper.

conditions of these experiments. These same experiments suggest a different function for the RNA–antigen complex but yield no clue as to its mode of action.

TRANSFER OF ANTIBODY FORMATION BY RNA FROM CELLS OF ANTIBODY-PRODUCING TISSUES (HORIZONTAL TRANSMISSION)

Closely related to the previously discussed reports on the recovery of highly immunogenic RNA–antigen complexes from blood and urine of immunized animals are the numerous findings of immunogenic RNA in spleens or lymph nodes of immune animals. In such experiments it is of course difficult to localize the cellular source of the active RNA fractions. From the data already discussed it seems reasonable to assume that RNA from such sources might well contain messenger-type material that activates the recipient's cells directly, and also RNA–antigen complexes that would act in a more indirect fashion. The recent papers by Konda *et al.* (1962), Cohen *et al.* (1965), Gerughty *et al.* (1966), and Mitsuhashi *et al.* (1967) are examples of the research efforts in this area. Of very special interest is the recent observation of Alexander (1968) that RNA extracted from lymph nodes of sheep that had been immunized against specific rat tumors caused upon injection into rats the production of antibody against the tumor and that such antibody possessed antigenic markers characteristic of sheep γ-globulin.

Another important observation is that of Friedman (1968), who has found RNase-sensitive immunogenic material in the spleen of mice that had been injected neonatally with *Shigella* even if the dose had been such that the spleen donors themselves were specifically tolerant. The state of delayed sensitivity to agents such as tuberculin has been transferred from immune to normal cells with the aid of RNA (Thor and Dray, 1968), as has the specific sensitization to transplantation antigens (Mannick and Egdahl, 1964). The report that RNA from "normal" lymphoid cells can convert "immune" cells to nonreactive cells (Mannick, 1964) warrants due attention in attempts to explain the mechanism of these apparent transformations. Finally, one must mention the extraordinary "transfer factor" (Lawrence, 1960), which has been shown to be dialyzable and highly resistant to degradation by all enzymes tested and which transfers specific, delayed-type hypersensitivity to nonreactive cells in a manner that suggests self-replication.

MORE CELLULAR INTERACTIONS: LYMPHOCYTES AFFECTING MACROPHAGES

Though rather tangential to our main theme, we feel that we should mention in passing that the wealth of information on cellular interactions in which the macrophage is the initiator and the lymphoid cell the executor is well balanced by a body of data which documents interactions in the reverse direction. It has been shown that in the course of the reaction between sensitized lymphoid cells and antigen one or more factors are released and that this material profoundly affects macrophages, which, as innocent bystanders, react by loss of their motility (David, 1968; Bloom and Bennett, 1968). Moving still one step further from our central theme we also should call attention to the well known phenomena of allergic reactions. In these reactions, contact of antigen with specifically reactive cells or with antibody triggers a complex series of events that culminates in the release of pharmacologically active substances from various bystander cells and affects a wide variety of target cells (Kolb and Granger, 1968).

A VIEW OF ANTIBODY SYNTHESIS INVOLVING INTERACTIONS BETWEEN THREE CELL TYPES

In recent years the role of the thymus and that of the bursa of Fabricius have been the subject of intensive investigation. We can deal with this material only in a most superficial manner, focusing our attention on direct evidence for cellular interactions in antibody formation. It has been reported that the ability of neonatally bursectomized birds to make normal levels of γ-globulin and to mount a normal antibody response against sheep red blood cells was restored by the implantation of pieces of bursal tissue. In contrast, the injection of dispersed bursal cells restored only the ability to make γ-globulin, not that to make specific antibodies to foreign red blood cells. One possible explanation that was advanced was that interaction between bursal cells was essential for the development of immunological competence (Cooper *et al.*, 1967). Similarly, Auerbach (1960) has called attention to the critical role of mesenchymal cells in bringing about lymphoid development in the embryonic thymus, and Hays (1967) has shown that surviving epidermal cells of a thymic transplant "condition" the recipient's lymphocytes which repopulate the transplant.

Claman *et al.* (1966) made the very important observation that X-irradiated mice would respond to sheep red blood cells much more effectively if given bone marrow and thymus cells than they did when given cells of one type only. This original observation has been elaborated extensively by others (Mitchell and Miller, 1968; Mosier, 1969; Miller and Mitchell, 1969) and has led to the conclusion that the thymus derived lymphoid cells act as "recognition" cells for a given antigen, but that antibody formation occurs in the bone-marrow derived lymphoid cells which act as nonspecific factory cells. Mosier and Coppelson's data (1968) on the induction of antibody formation to sheep red cells in cultures of normal spleen cells have led to the suggestion that three cell types are required, namely, one class of cells located in the fraction that adheres to the culture container, and cells of two additional classes among the nonadherent population. Since the number of competent cells in each population appeared to be limited, the authors feel that the response to the test antigen may involve interaction between at least two and possibly three antigen-specific cells. Talmage *et al.* (1969) have arrived at similar conclusions. They suggest that adherent cells preferentially obtained from spleens of X-irradiated animals that had been repopulated with bone marrow cells react with the antigen and, in addition, function so as to attract and bring together representatives of the two classes of nonadherent lymphoid cells, those that are derived from bone marrow and those others that originated in or passed through the thymus. In one model the thymic cell is a memory or recognition cell which makes specific messenger RNA and passes this to the bone marrow-derived potential antibody producer. In an alternate model (credited to E. Lennox), the thymic cell is considered to be one with specific receptors for the antigen which functions so as to enhance the reaction between the antigen-loaded adherent cell and the antibody-containing lymphoid cell of bone marrow origin which is destined to proliferate and to make antibody. The separation of cells into adherent and nonadherent classes is a technique open to some criticism since the protein content of the medium will profoundly affect the distribution of cells in these two "compartments" (Shortman, 1966), and because it can be shown that the adherent cells may contain a good number of antibody producers (Holub, 1967; Plotz and Talal, 1967).

A recent finding of Kölsch (1968) deserves attention while the role of thymus or thymus derived (or passaged) lymphocytes is under consideration. Working with mice he found that macrophages carrying a labeled antigen accumulated in significant numbers in the thymus within 2–4 hours after injection. It seems possible, therefore, that in the intact animal the initial contact of the two nonproducers of antibody in the three-cell model may occur in the thymus.

DISCUSSION, SPECULATION, AND CONCLUSIONS

It has been a relatively easy matter to compile the facts, hard and soft, that have been set forth in the preceding pages. But progress in science requires more than facts, some of which may be tomorrow's artifacts, and calls for concepts, hypotheses, and theories. We feel that the *in vitro* studies cited have firmly established that cellular interactions are of importance and probably essential in the induction of antibody formation, and that results of *in vivo* studies support this concept strongly. We also believe that there is sufficient evidence from both *in vitro* and *in vivo* data to establish RNA–antigen complexes as an important though not necessarily sole mediator in immunogenesis.

Before we go into a discussion of these subjects, it might be well to state our view of the transfer of informational RNA and its possible role. Our data on the transfer of informational RNA from macrophages to lymphoid cells that respond by producing antibody of the macrophage donor's allotype are most likely a result of the existence of a small number of macrophages which produce trace amounts of the specific antibody in question. That it is only a small number of such cells we deduce from two findings. (1) The addition of one bacteriophage particle to 1000 macrophages yields RNA of the same specific biological activity as does the addition of much larger amounts of phage (10^5 particles), an indication that "saturation" is reached with the small dose. (2) Opsonization of the phage or nonspecific stimulation of the phagocytic activity of the macrophages results in greatly enhanced phagocytic activity, yet again no further increase in RNA activity is observed. Our estimate of no more than one "competent" macrophage per thousand is compatible with the deduction of Mosier and Coppleson (1968) that in their system there are no more than 1–10 responsive cells per million spleen cells. The

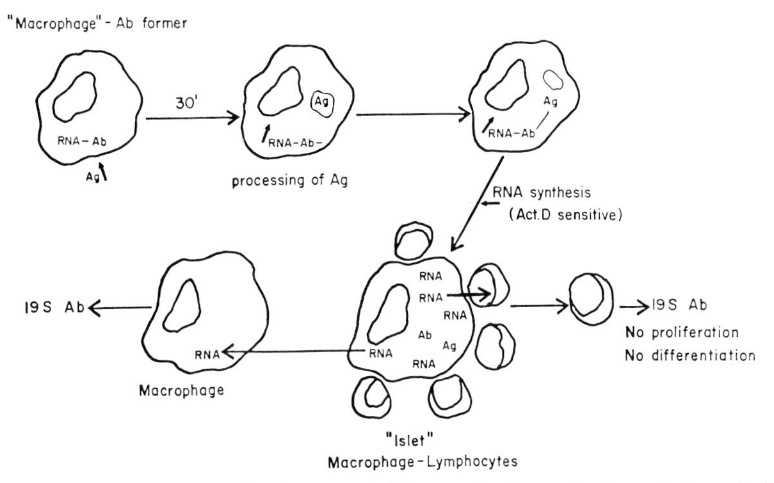

FIG. 1. Macrophages with the potential for antibody production. Antigen is ingested or adsorbed, a step probably mediated by specific antibody produced by and attached to the cell. A specific messenger-type RNA is formed which results in the synthesis of 19 S antibody either in the same cell or, upon transfer to adherent lymphoid cells, in the latter cells. The "islet" shown provides for amplification of the antibody response in the absence of cell proliferation or differentiation. From Fishman (1969).

evidence for antibody production by some macrophages has already been reviewed in a preceding passage. We suggest that such cells preferentially capture the antigen through antibody bound to their membranes, and that they respond to this event by the synthesis of relevant messenger-type RNA.

If this model were applicable to the intact animal, one could envisage passage of informational RNA through cytoplasmic bridges or lymphoid uropods from macrophages into lymphoid cells, a process which would lead to amplification of an antibody response that would be limited in time to the functional survival of the RNA. Such antibody formation would presumably leave no "memory" and would involve no cell replication. This situation has been observed *in vivo* (Tannenberg and Malaviya, 1968) and *in vitro* (Bussard, 1966).

In offering this suggestion we are aware of several possible objections of which the major one rests on the fact that mRNA of the 8–10 S class may be too small to code for an entire L chain (mol. wt. = 20,000) or an entire H polypeptide chain (mol. wt. = 50,000). Biochemical evidence suggests that these chains are synthesized as units, on separate polyribosomes (Askonas and Williamson, 1967; Lennox *et al.*, 1967), starting at their respective amino terminal ends (i.e., the termini of the variable portions). Nevertheless, analyses of amino acid sequences of immunoglobulins suggest genetic mechanisms that are difficult to reconcile with a simplistic one gene-one messenger-one polypeptide view (Hood *et al.*, 1967), and the active RNA fraction under discussion could well code for the variable portions of the chains, the portions which—in rabbit antibody molecules—contain both the allotypic markers and the antibody combining sites (Kosland, 1967). While we have as yet no evidence from our work that informational RNA could transfer 7 S antibody formation from cell to cell, this should be possible with the aid of extracts from IgG-producing cells (Herscowitz and Stelos, 1968).

Proceeding to the role of the RNA-antigen complex in the induction of antibody formation, we would call attention again to the outstanding properties of potent immunogenicity ("superantigen") and the impairment of this by enzymatic degradation of either of the two components. This lability may indicate no more than that the RNA stabilizes and preserves antigenic determinants, an RNA function for which there admittedly exists no known precedent. Alternatively, one could imagine that the RNA part may be the key which opens the antibody producing cell to the antigen which then may act as a derepressor (Szilard, 1960) or as a template (Haurowitz, 1969). In contrast, Braun (1968) has suggested that the antigen part of the complex may play the role of the vehicle that brings the complex to potential antibody producers, which recognize and capture it by means of specific receptors (cell-bound antibody), and that the function of the RNA is that of a derepressor. An intriguing recent suggestion for the role of macrophage RNA is that of Harshman *et al.* (1969), who believe that this RNA may contain the initiator codon for antibody synthesis and thus triggers antibody synthesis in competent cells. These arguments suffer from the uncertainty as to whether antigen must indeed enter the cell that is destined to make antibody, a problem that remains the subject of lively controversy (Nossal, 1966a; Nossal *et al.*, 1967).

There are those who consider antigen held on the surface of the macrophage to be the factor relevant to immunogenesis (Askonas *et al.*, 1968; Mitchison, 1969). It is by no means clear in what form such antigen exerts its activity; Mitchison (1969) considers it quali-

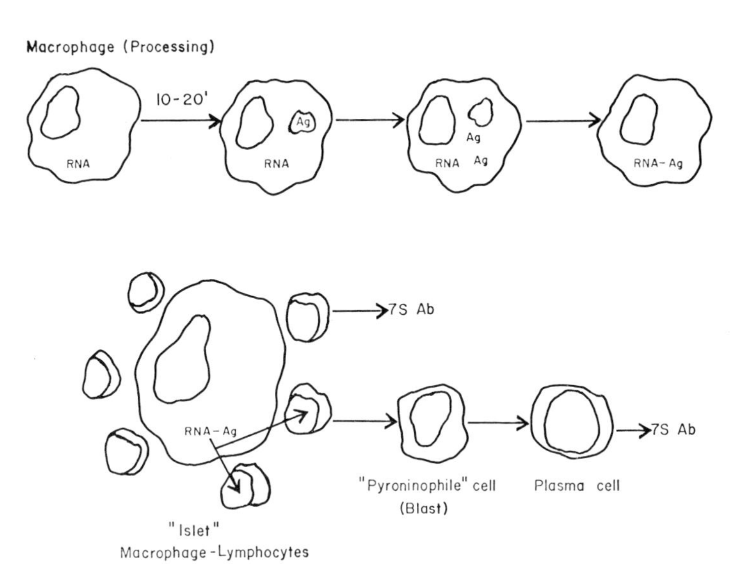

FIG. 2. Macrophages that process the antigen into an RNA-antigen complex. The ingested antigen is solubilized and bound to preexistent RNA. The resulting complex is transferred to lymphoid cells that adhere to the macrophage. These cells, in turn, proliferate, differentiate, and produce antibody. From Fishman (1969).

tatively and quantitatively different from free antigen. It also remains to be seen why such antigen would resist enzymatic degradation and what would prevent its masking by antibody if it were to persist. Could it possibly be held, in a stabilized form, by membrane-associated RNA (Weiss and Mayhew, 1966)? Other, less specific roles of the macrophage in immunogenesis have also been considered. Askonas *et al.* (1968) suggest, as one of several possibilities, that in their interaction with lymphoid cells the macrophages may exert a regulatory role such as supplying the stimulus for division or differentiation. *In vitro* studies have furnished some evidence that cells adherent to glass may release into the medium as yet unidentified factors that play a role in the induction of antibody formation (Dutton, 1969, personal communication).

If the effective RNA–antigen complex exists intracellularly, one must assume that it has escaped the lysosome-phagosome pathway toward total destruction (Ehrenreich and Cohn, 1967). That such sanctuaries exist, possibly in a special, functional subclass of macrophages, has been suggested by the data presented. In a recent publication Kölsch and Mitchison (1968) have discussed such "compartments." We would suggest that partial degradation and enzymatic coupling of the antigen fragment to RNA occurs in extralysosomal

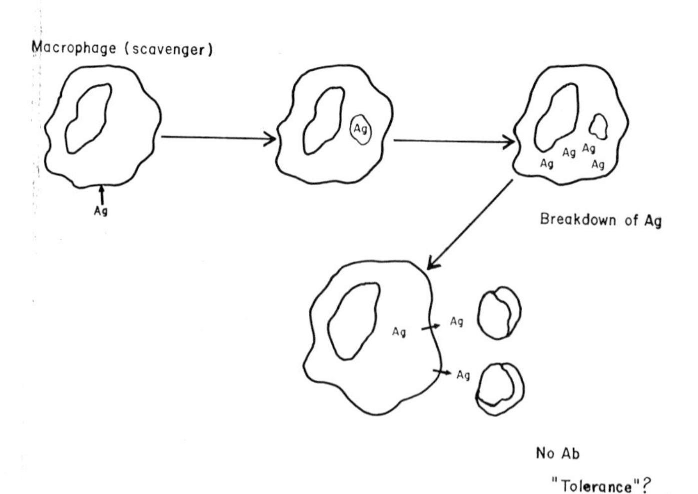

FIG. 3. "Classical" macrophages. The antigen enters the phagosome-lysosome pathway toward complete digestion. Partially digested antigen may play a role in the induction of immunological tolerance. From Fishman (1969).

parts of the cytoplasm of certain macrophages, and that such complexes may in the course of events appear on the surface of the cell in question. Since peritoneal but not alveolar macrophages are active in our system, evidence for the functional heterogeneity of macrophages is at hand. That macrophages may arise from precursor cells either in the marrow or in the thoracic duct lymph has been demonstrated (Boak *et al.*, 1968), but the dual origin of this cell is not universally accepted (Z. A. Cohn, 1968). There also is a suggestion that the macrophage instrumental in immunogenesis can be differentiated from the phagocytic-destructive cell by the greater radiosensitivity of the precursor of the former cell (Feldman, 1968; Mitchison, 1969). It is to be hoped that further work will help toward the identification of the macrophages that are scavengers and those that are productive of the RNA–antigen complex, and that such work will dispel the confusion which probably has its root in the functional and genealogical heterogeneity of these cells.

While the more conspicuous activity of a population of macrophages is that of ingestion and degradation of the antigen, a feature which has led Burnet (1959) to assign to them the role of protecting lymphoid cells against injurious exposure to massive antigen doses, their ability to link antigen fragments to RNA into a highly immunogenic and possibly quite stable form would suggest another, and quite opposite function, namely that of an alma mater of antibody formation. It remains for future work to resolve these problems, but there remains no doubt that cellular interactions play an important role in immunogenesis. It is their significance and their mode of interaction that require further elucidation.

ACKNOWLEDGMENT

This work was supported by Grant AI-06899 from the National Institute of Allergy and Infectious Diseases.

Figures 1–3 are reproduced by permission of Annual Reviews, Inc. which we gratefully acknowledge. They have appeared in Volume 23 of the Annual Review of Microbiology as part of an article on "Induction of Antibodies in Vitro" by M. Fishman.

REFERENCES

Ada, G. L., Lang, P. G., and Plymin, G. (1968). Antigen in tissues. V. Effect of endotoxin on the fate of, and on the immune response to, serum albumin and to albumin-antibody complexes. *Immunology* **14,** 825–836.

Adler, F. L., Fishman, M. and Dray, S. (1966). Antibody formation initiated in vitro. III. Antibody formation and allotypic specificity directed by ribonucleic acid from peritoneal exudate cells. *J. Immunol.* **97,** 554–558.

ALEXANDER, P. (1968). Immunotherapy of cancer: Experiments with primary tumours and syngeneic tumour grafts. *Progr. Exptl. Tumor Res.* **10,** 22–71.

ARGYRIS, B. F. (1968). Role of macrophages in immunological maturation. *J. Exptl. Med.* **128,** 459–467.

ASKONAS, B. A., and RHODES, J. M. (1965). Immunogenicity of antigen-containing ribonucleic acid preparations from macrophages. *Nature* **205,** 470–474.

ASKONAS, B. A., and WILLIAMSON, A. R. (1967). Biosynthesis and assembly of immunoglobulin G. *Cold Spring Harbor Symp. Quant. Biol.* **32,** 223–231.

ASKONAS, B. A., AUZINS, I., and UNANUE, E. R. (1968). Role of macrophages in the immune response. *Bull. Soc. Chim. Biol.* **50,** 1113–1128.

AUERBACH, R. (1960). Morphogenetic interaction in the development of the mouse thymus gland. *Develop. Biol.* **2,** 271–284.

BENDINELLI, M. (1968). Haemolytic plaque formation by mouse peritoneal cells, and the effect on it of Friend virus infection. *Immunology* **14,** 837–850.

BISHOP, D. C., PISCIOTTA, A. V., and ABRAMOFF, P. (1967). Synthesis of normal and "immunogenic RNA" in peritoneal macrophage cells. *J. Immunol.* **99,** 751–759.

BLOOM, B. R., and BENNETT, B. (1968). Migration inhibitory factor associated with delayed-type hypersensitivity. *Federation Proc.* **27,** 13–15.

BOAK, J. L., CHRISTIE, G. H., FORD, W. L., and HOWARD, J. G. (1968). Pathways in the development of liver macrophages: Alternative precursors contained in populations of lymphocytes and bone-marrow cells. *Proc. Roy. Soc.* **B169,** 307–327.

BRAUN, W. (1968). *In* "Nucleic Acids in Immunology" (O. J. Plescia and W. Braun, eds.), Discussion, pp. 535–537. Springer, New York.

BURNET, F. M. (1959). Auto-immune disease. I. Modern immunological concepts. *Brit. Med. J.* **2,** 645–650.

BUSSARD, A. E. (1966). Antibody formation in non-immune mouse peritoneal cells after incubation in gum containing antigen. *Science* **153,** 887–888.

CAMPBELL, D. H., and GARVEY, J. S. (1963). Nature of the retained antigen and its role in immune mechanisms. *Advan. Immunol.* **3,** 261–313.

CARREL, A., and INGEBRIGTSEN, R. (1912). The production of antibodies by tissues living outside of the organism. *J. Exptl. Med.* **15,** 287–291.

CLAMAN, H. N., CHAPERON, E. A., and TRIPLETT, R. F. (1966). Thymus-marrow cell combinations. Synergism in antibody production. *Proc. Soc. Exptl. Biol. Med.* **122,** 1167–1171.

COHEN, E. P., and RASKA, K., JR. (1967). Antigen-unique species of RNA in peritoneal cells of mice which do not adhere to glass. *Cold Spring Harbor Symp. Quant. Biol.* **32,** 349–351.

COHEN, E. P., NEWCOMB, R. W., and CROSBY, L. K. (1965). Conversion of nonimmune spleen cells to antibody-forming cells by RNA: Strain specificity of the response. *J. Immunol.* **95,** 583–590.

COHN, M. (1967). Antibody synthesis: The take home lesson. *In* "Gammaglobulins; Structure and Control of Biosynthesis" (J. Killander, ed.), pp. 615–643. Wiley (Interscience), New York.

COHN, Z. A. (1968). The differentiation of macrophages. *Symp. Intern. Soc. Cell Biol.* **7,** 101–110.

COOPER, M. D., GABRIELSEN, E., and GOOD, R. A. (1967). The development, differentiation, and function of the immunoglobulin production system. *In* "Ontogeny of Immunity" (R. T. Smith, R. A. Good, and P. A. Miescher, eds.), pp. 122–132. Univ. of Florida Press, Gainesville, Florida.

DAVID, J. R. (1968). Macrophage migration. *Federation Proc.* **27,** 6–12.

DRESSER, D. W., and MITCHISON, N. A. (1968). The mechanism of immunological paralysis. *Advan. Immunol.* **8,** 129–181.

DUTTON, R. W., and MISHELL, R. I. (1967). Cellular events in the immune response. The in vitro response of normal spleen cells to erythrocyte antigens. *Cold Spring Harbor Symp. Quant. Biol.* **32,** 407–414.

EHRENREICH, B. A., and COHN, Z. A. (1967). The uptake and digestion of iodinated human serum albumin by macrophages in vitro. *J. Exptl. Med.* **126,** 941–958.

EHRICH, W. E. (1956). *In* "Handbuch der allgemeinen Pathologie" (F. Buchner, E. Letterer, and F. Roulet, eds.), Vol. 7, Part 1, p. 137. Springer, Berlin.

FELDMAN, M. (1968). The immunogenic function of macrophages. *Symp. Intern. Soc. Cell Biol.* **7,** 43–48.

FISHMAN, M. (1959). Antibody formation in tissue culture. *Nature* **183,** 1200–1201.

FISHMAN, M. (1961). Antibody formation in vitro. *J. Exptl. Med.* **114,** 837–856.

FISHMAN, M. (1969). Induction of antibodies in vitro. *Ann. Rev. Microbiol.* **23,** 199–222.

FISHMAN, M., and ADLER, F. L. (1963). Antibody formation initiated in vitro. II. Antibody synthesis in X-irradiated recipients of diffusion chambers containing nucleic acid derived from macrophages incubated with antigen. *J. Exptl. Med.* **117,** 595–602.

FISHMAN, M., and ADLER, F. L. (1967). The role of macrophage-RNA in the immune response. *Cold Spring Harbor Symp. Quant. Biol.* **32,** 343–347.

FISHMAN, M., ADLER, F. L., and HOLUB, M. (1968). Antibody formation initiated in vitro with RNA and RNA-antigen complexes. *In* "Nucleic Acids in Immunology" (O. J. Plescia and W. Braun, eds.), pp. 439–446. Springer, New York.

FISHMAN, M., ADLER, F. L., VAN ROOD, J. J., and BINET, J. L. (1965a). Macrophage involvement in antibody formation in vitro. *Proc. 4th Intern. Symp. (Reticuloendothel. System) Otsu RES Kyoto, Japan, 1964,* pp. 229–237.

FISHMAN, M., VAN ROOD, J. J., and ADLER, F. L. (1965b). The initiation of antibody formation by ribonucleic acid from specifically stimulated macrophages. *In* "Molecular and Cellular Basis of Antibody Formation" (J. Sterzl with co-workers, eds.), pp. 491–501. Publ. House, Czechoslovak Acad. Sci., Prague.

FORD, W. L., GOWANS, J. L., and MCCULLAGH, P. J. (1966). The origin and function of lymphocytes. *Ciba Found. Symp. Thymus Exptl. Clin. Studies,* p. 58. Little, Brown, Boston, Massachusetts.

FREI, P. C., BENACERRAF, B., and THORBECKE, G. J. (1965). Phagocytosis of the antigen, a crucial step in the induction of the primary response. *Proc. Natl. Acad. Sci. U.S.* **53,** 20–23.

FRIEDMAN, H. (1968). The nature of immunogenic RNA-antigen complexes in immune and tolerant mice. *In* "Nucleic Acids in Immunology" (O. J. Plescia and W. Braun, eds.), pp. 505–526. Springer, New York.

FRIEDMAN, H. P., STAVITSKY, A. B., and SOLOMON, J. M. (1965). Induction in vitro of antibodies to phage T2: Antigens in the RNA extract employed. *Science* **149,** 1106–1107.

GARVEY, J. S., CAMPBELL, D. H., and DAS, M. Z. (1967). Urinary excretion of foreign antigen and RNA following primary and secondary injections of antigens. *J. Exptl. Med.* **125,** 111–126.

GERSHON, H., and FELDMAN, M. (1968). Studies on the immune reconstitution of sublethally irradiated mice by peritoneal macrophages. *Immunology* **15,** 827–835.

GERUGHTY, R. M., ROSENAU, W., and MOON, H. D. (1966). In vitro transfer of immunity by ribosomes. *J. Immunol.* **97,** 700–708.

GOTTLIEB, A. A. (1968). The antigen-RNA complex of macrophages. *In* "Nucleic Acids in Immunology" (O. J. Plescia and W. Braun, eds.), pp. 471–486. Springer, New York.

GOTTLIEB, A. A., GLISIN, V. R., and DOTY, P. (1967). Studies on macrophage RNA involved in antibody production. *Proc. Natl. Acad. Sci. U.S.* **57,** 1849–1856.

HALAC, E., RIFE, U., and RINALDINI, L. M. (1965). Antigenically induced RNA changes in macrophages and lymph nodes. *In* "Symposium on the Mutational Process, Genetic Variation in Somatic Cells" (J. Klein, M. Vistiskova, and V. Zeleny, eds.), pp. 249–253. Publ. House, Czechoslovak Acad. Sci., Prague.

HARRIS, T. N., HUMMELER, K., and HARRIS, S. (1966). Electron microscopic observations on antibody producing lymph node cells. *J. Exptl. Med.* **123,** 161–172.

HARSHMAN, S., DUKE, L. J., and SIX, H. (1969). Mechanism of antibody biosynthesis. I. Isolation and physical chemical properties of the retained antigen. *Immunochemistry* **6,** 175–188.

HAUROWITZ, F. (1969). Struktur und Wirkungsweise der Antikörper. *Naturwissenschaften* **56,** 189–194.

HAYS, E. F. (1967). The effects of allografts of thymic epithelial reticular cells on the lymphoid tissue of neonatally thymectomized mice. *Blood* **29,** 29–40.

HERSCOWITZ, H. B., and STELOS, P. (1968). The induction of bacteriophage neutralizing substances by RNA. *Federation Proc.* **27,** 317 (Abstract).

HOLUB, M. (1967). The nature of the activated small lymphocyte. *In* "The Lymphocytes in Immunology and Haemopoiesis" (J. M. Yoffey and F. C. Courtice, eds.), pp. 46–55. Williams & Wilkins, Baltimore, Maryland.

HOOD, L., GRAY, W. R., SANDERS, B. G., and DREYER, W. J. (1967). Light chain evolution. *Cold Spring Harbor Symp. Quant. Biol.* **32,** 133–146.

HUMPHREY, J. H., and FRANK, M. M. (1967). The localization of non-microbial antigens in the draining lymph nodes of tolerant, normal and primed rabbits. *Immunology* **13,** 87–100.

JAFFÉ, R. H. (1931). The reticulo-endothelial system in immunity. *Physiol. Rev.* **11,** 277–327.

KOLB, W. P., and GRANGER, G. A. (1968). Lymphocyte in vitro cytotoxicity: Characterization of human lymphotoxin. *Proc. Natl. Acad. Sci. U.S.* **61,** 1250–1255.

KÖLSCH, E. (1968). Migration of macrophages carrying antigen into the thymus. *Experientia* **24,** 951–953.

KÖLSCH, E., and MITCHISON, N. A. (1968). The subcellular distribution of antigen in macrophages. *J. Exptl. Med.* **128,** 1059–1079.

KONDA, S., NORO, Y., SAWAI, Y., and TASHIRO, Y. (1962). A role of lymphocytes in antibody formation: Antibody formation promoting function of ribonucleic acid isolated from lymphocytes. *Acta Haematol. Japon.* **25,** 159–168.

KOSHLAND, M. E. (1967). Location of specificity and allotypic amino acid residues in antibody Fd fragments. *Cold Spring Harbor Symp. Quant. Biol.* **32,** 119–127.

LAWRENCE, H. S. (1960). Some biological and immunological properties of transfer factor. *Ciba Found. Symp. Cellular Aspects Immunity*, pp. 243–271. Little, Brown, Boston, Massachusetts.

LENNOX, E. S., KNOPF, P. M., MUNRO, A. J., and PARKHOUSE, R. M. E. (1967). A search for biosynthetic subunits of light and heavy chains of immunoglobulins. *Cold Spring Harbor Symp. Quant. Biol.* **32,** 249–254.

LITT, M. (1967). Studies of the latent period. I. Primary antibody in guinea pig lymph nodes 7½ minutes after introduction of chicken erythrocytes. *Cold Spring Harbor Symp. Quant. Biol.* **32,** 477–492.

McMASTER, P. D. (1961). Antibody formation. *In* "The Cell" (J. Brachet and A. E. Mirsky, eds.), Vol. 5, pp. 323–404. Academic Press, New York.

MANNICK, J. A. (1964). Inhibition by RNA of the transfer reaction following homograft sensitization. *J. Clin. Invest.* **43,** 740–750.

MANNICK, J. A., and EGDAHL, R. H. (1964). Transfer of heightened immunity to skin homografts by lymphoid RNA. *J. Clin. Invest.* **43,** 2166–2177.

MARBROOK, J. (1967). Primary immune response in cultures of spleen cells. *Lancet* **II,** 1279–1281.

MILLER, J. F. A. P., and MITCHELL, G. F. (1969). Cell to cell interaction in the immune response. *Transplantion Proc.* **1,** 535–538.

MITCHELL, G. F., and MILLER, J. F. A. P. (1968). Immunological activity of thymus and thoracic duct lymphocytes. *Proc. Natl. Acad. Sci. U.S.* **59,** 296–303.

MITCHISON, N. A. (1968). Immunological paralysis as a dosage phenomenon. *In* "Regulation of the Antibody Response" (B. Cinader, ed.), pp. 54–67. Thomas, Springfield, Illinois.

MITCHISON, N. A. (1969). The immunogenic capacity of antigen taken up by peritoneal exudate cells. *Immunology* **16,** 1–14.

MITSUHASHI, S., SAITO, K., OSAWA, N., and KURASHIGE, S. (1967). Experimental Salmonellosis. XI. Induction of cellular immunity and formation of antibody by transfer agent of mouse mononuclear phagocytes. *J. Bacteriol.* **94,** 907–913.

MOSIER, D. E. (1967). A requirement for two cell types for antibody formation in vitro. *Science* **158,** 1573–1575.

MOSIER, D. E. (1969). Cell interactions in the primary immune response in vitro: A requirement for specific cell clusters. *J. Exptl. Med.* **129,** 351–362.

MOSIER, D. E., and COHEN, E. P. (1968). Induction and rapid expression of an immune response in vitro. *Nature* **219,** 969–970.

MOSIER, D. E., and COPPLESON, L. W. (1968). A three-cell interaction required for the induction of the primary immune response in vitro. *Proc. Natl. Acad. Sci. U.S.* **61,** 542–547.

NOLTENIUS, H., and CHAHIN, M. (1969). Further evidence concerning macrophages producing 19S-antibody in mice. *Experientia* **25,** 401.

NOSSAL, G. J. V. (1966a). Immunological tolerance: A new model system for low zone induction. *Ann. N.Y. Acad. Sci.* **129,** 822–833.

NOSSAL, G. J. V. (1966b). Antibody production in tissue culture. *In* "Cells and Tissues in Culture" (E. N. Willmer, ed.), pp. 317–350. Academic Press, New York.

NOSSAL, G. J. V., ADA, G. L., and AUSTIN, C. M. (1964). Antigens in immunity. II. Immunogenic properties of flagella, polymerized flagellin and flagellin in the primary response. *Australian J. Exptl. Biol. Med. Sci.* **42,** 283–294.

NOSSAL, G. J. V., WILLIAMS, G. M., and AUSTIN, C. M. (1967). Antigens in immunity. XIII. The antigen content of single antibody-forming cells early in primary and secondary immune responses. *Australian J. Exptl. Biol. Med. Sci.* **45,** 581–594.

PANIJEL, J., and CAYEUX, P. (1968). Immunosuppressive effects of macrophage antiserum. *Immunology* **14,** 769–780.

PINCHUCK, P., MAURER, P. H., FISHMAN, M., and ADLER, F. L. (1968). Antibody formation: Initiation in "non-responder" mice by macrophage synthetic polypeptide RNA. *Science* **160,** 194–195.

PLOTZ, P. H., and TALAL, N. (1967). Fractionation of splenic antibody-forming cells on glass bead columns. *J. Immunol.* **99,** 1236–1242.

ROELANTS, G. E., and GOODMAN, J. W. (1968). Immunochemical studies on the poly-γ-D-glutamyl capsule of *B. anthracis*. IV. The association with peritoneal exudate cell ribonucleic acid of the polypeptide in immunogenic and nonimmunogenic forms. *Biochemistry* **7,** 1432–1440.

SCHOENBERG, M. D., MUMAW, V. R., MOORE, R. D., and WEISBERGER, A. S. (1964). Cytoplasmic interaction between macrophages and lymphocytic cells in antibody synthesis. *Science* **143,** 964–965.

SHANDS, J. W., JR. (1967). The immunological role of the macrophage. *In* "Modern Trends in Immunology" (R. Cruickshank and D. M. Weir, eds.), Vol. 2, pp. 86–118. Appleton-Century-Crofts, New York.

SHARP, J. A., and BURWELL, R. G. (1960). Interaction ("Peripolesis") of macrophages and lymphocytes after skin homografting or challenge with soluble antigens. *Nature* **188,** 474–475.

SHORTMAN, K. (1966). The separation of different cell classes from lymphoid organs. I. The use of glass bead columns to separate small lymphocytes, remove damaged cells and fractionate cell suspensions. *Australian J. Exptl. Biol. Med. Sci.* **44,** 271–286.

STRAUS, W. (1967). Lysosomes, phagosomes and related particles. *In* "Enzyme Cytology" (D. B. Rooedyn, ed.), pp. 239–319. Academic Press, New York.

SULITZEANU, D. (1968). Affinity of antigen for white cells and its relation to the induction of antibody formation. *Bacteriol. Rev.* **32,** 404–424.

SZILARD, L. (1960). The molecular basis of antibody formation. *Proc. Natl. Acad. Sci. U.S.* **46,** 293–302.

TALMAGE, D. W., RADOVICH, J., and HEMMINGSEN, H. (1969). Cell interaction in antibody synthesis. *J. Allergy* **43,** 323–335.

TANNENBERG, W. J. K., and MALAVIYA, A. N. (1968). The life cycle of antibody-forming cells. I. The generation time of 19S hemolytic plaque-forming cells during the primary and secondary responses. *J. Exptl. Med.* **128,** 895–925.

THOR, D. E., and DRAY, S. (1968). The cell-migration-inhibition correlate of delayed hypersensitivity. Conversion of human nonsensitive lymph node cells to sensitive cells with an RNA extract. *J. Immunol.* **101,** 469–480.

UNANUE, E. R., and ASKONAS, B. A. (1968). The immune response of mice to antigen in macrophages. *Immunology* **15,** 287–296.

WEIGLE, W. O., and GOLUB, E. S. (1967). Kinetics of the establishment of immunological unresponsiveness to serum protein antigens. *Cold Spring Harbor Symp. Quant. Biol.* **32,** 555–558.

WEISS, L., and MAYHEW, E. (1966). The presence of ribonucleic acid within the peripheral zones of two types of mammalian cells. *J. Cell. Physiol.* **68,** 345–360.

DEVELOPMENTAL BIOLOGY SUPPLEMENT 3, 133–150 (1969)

Nervous and Hormonal Communication in Insect Development

CARROLL M. WILLIAMS

Department of Biology, Harvard University, Cambridge, Massachusetts 02138

THE GENETIC CONSTRUCTION MANUAL

A developing organism is an expanding community of cellular lives, each centering in itself and yet each presupposing the whole. The growth and differentiation of this community takes place in a coordinated manner according to a detailed "construction manual" inherited from the preceding generation.

In the case of the higher insects, the construction manual presents certain novelties in that it is subdivided into three successive chapters. The first chapter gives instructions for turning the egg into a larva. Days or weeks later, the second chapter tells how to transform the larva into an essentially new organism, the pupa. Then, within the closed system of the pupa, the third and final chapter prescribes the reworking of the cells and tissues to form the adult. This analogy serves to emphasize that the more advanced forms of metamorphosis involve the derepression and acting-out of what is little short of successive batches of genetic information. We may think of the genome as being subdivided into three different "gene sets" corresponding to the successive chapters in the construction manual.

The construction manual, to continue the analogy, is read by the individual cells, which even in the smallest insects are numbered in the tens of thousands. In a favorite experimental animal, the Cecropia silkworm (*Hyalophora cecropia*), the pupa contains about a billion cells. Manifestly, the metamorphosis of the insect as a whole is the outcome of a mosaic of metamorphoses at the cellular level. And it is worth inquiring how all these cells are coordinated in the playback of their successive sets of genes.

There is a wealth of evidence to document the answer to this question. In insects as in human beings the coordination of the cellular community is the job of the two great systems of communication—the nervous system and the endocrine system.

The actions and interactions of these channels of communication during the growth and metamorphosis of insects are considered herein. In more ways than one, this presentation constitutes variations on a theme discussed before this Society some 21 years ago (Williams, 1948).

MORPHOGENESIS AND THE NERVOUS SYSTEM

The neural coordination of development can be examined in a direct manner by excising the central nervous system. This audacious surgery has been carried out on silkworm pupae (*Hyalophora cecropia* and *Antheraea polyphemus*) in the following manner:

A pupa is deeply anesthetized with carbon dioxide and its facial integument cut away with a scalpel. Through the facial opening one removes the brain, the subesophageal ganglion, and the three thoracic ganglia. The opening is covered with a disk of plastic (cut from a plastic coverslip); the disk is sealed in place with melted wax. The tip of the abdomen is then cut away, and as much blood as possible is drained into a vial containing a few crystals of the tyrosinase inhibitor phenylthiourea. Through the posterior opening one can see the chain of abdominal ganglia extending along the midventral line. By means of watchmaker's forceps, the bilateral nerves and tracheae are broken and the chain of abdominal ganglia is removed. The blood is then returned to the abdomen, all air expelled, and the opening sealed with a plastic window. Preparations of this type often survive for many months, as indicated by the beating of the heart. For our present purposes, what can we say about the career of such impoverished pupae?

The answer is unambiguous: no further development takes place except for the healing of the integumentary wounds. Yet, strange to say, it is not difficult to "wake up" these pupae and cause them to undergo adult development. All one has to do is to implant into the head, thorax, or abdomen a "loose" brain obtained from a chilled pupa of the same or even a different species of silkworm. Two to three weeks later the formation of the adult is initiated. After three additional weeks, development is complete, as signalled by the resorption of the molting fluid. When the pupal exuvia is peeled away, one exposes a flaccid moth.

As illustrated in Fig. 1, the moth appears to have undergone full and complete development. The compound eyes, the legs, the featherlike antennae, the genitalia, and all other integumentary structures cannot be distinguished from the corresponding organs of a normal moth. This perfection of integumentary development extends to the finest details of the pigmentation and patternization of the adult wings.

However, one cannot fail to be amazed when the moth is cut open and inspected. No new muscles have formed! Particularly impressive is the status of the thorax, where one can detect only traces of the numerous flight and leg muscles so characteristic of normal moths (Williams and Schneiderman, 1952).

A further surprise is the presence and complete metamorphosis of all other internal organs: the gut, the fat-body, the Malpighian tubules, the tracheal system, the gonads, and the accessory sex organs. In the case of females the abdomens contain mature ova, and in males, the testes and vasa deferentia are packed with spermatozoa.

From experiments of this type we learn that for all tissues except somatic muscle the resumption of development requires a non-species-specific factor that can be supplied by the implantation of a brain lacking all its nervous connections. We learn, moreover, that the differentiation of muscles requires some additional influence derived from the nervous system.

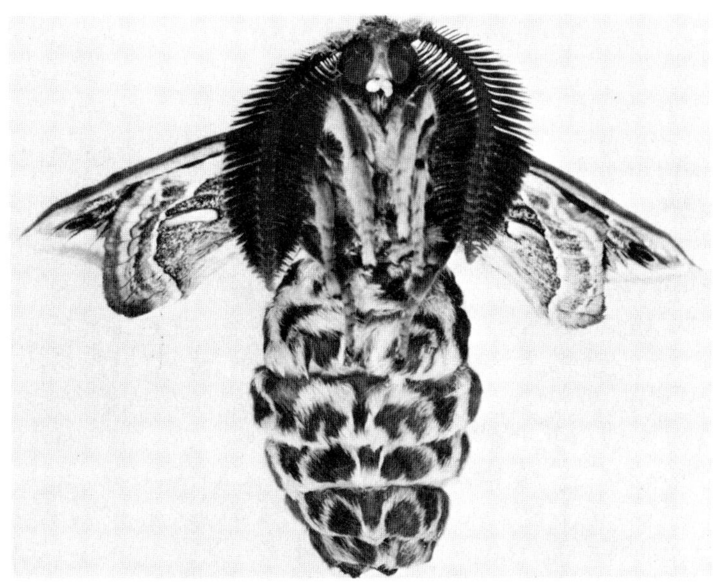

FIG. 1. Male moth (*Hyalophora cecropia*) derived from a pupa from which the entire central nervous system had been excised. Development was provoked by the implantation of an active pupal brain into the tip of the pupal abdomen.

The nature of this influence has been studied in utmost detail by Nüesch (1952, 1968). Nüesch found that it was not necessary to extirpate the nervous system: muscle formation can be blocked merely by cutting the nerves that pass from the ganglia to the muscle anlagen. When individual nerves were transected, certain predictable muscles failed to differentiate during adult development. When the denervation was postponed for one or more days after the initiation of development, the influence of the nervous system was maximal during the first few days and then declined. That being so, the longer the denervation was postponed, the greater the size attained by the corresponding muscles.

These findings strongly argue that the influence prerequisite for muscle formation is a trophic factor which must be delivered directly to the muscle anlagen by the motor nerves which innervate them.

THE BRAIN HORMONE

In our progress to this point we have identified two forms of communication between the central nervous system and the cells and tissues of the developing insect. The trophic factor is aimed exclusively at the muscle anlagen. By contrast, the brain communicates an endocrine signal which is somehow able to mobilize a developmental response in all tissues except muscle.

The endocrine function of the brain is an old story to students of insect physiology, having been discovered by Kopeć in 1922. The hormone, as first demonstrated by Wigglesworth (1940), is secreted by certain of the brain's neurosecretory cells. This latter finding is of great historical interest because it constituted the first experimental proof of an endocrine function for the neurosecretory cells of any organism.

In the pupal brain of the Cecropia silkworm there are 40 neurosecretory cell bodies arranged as follows (Fig. 2): a pair of medial groups containing a total of 22 cell bodies of four distinguishable types; a pair of lateral groups containing a total of 14 cell bodies of two distinct types; and a pair of posterior groups containing a total of 4 cells of a single type (Herman and Gilbert, 1965). So, in terms of their location and staining reactions, no less than seven types of neurosecretory cells can be identified in the brain of the Cecropia silkworm. The cytological picture therefore supports the ever-increasing physiological evidence that the insect brain is the source of a multiplicity of neurosecretions. However, in the present account,

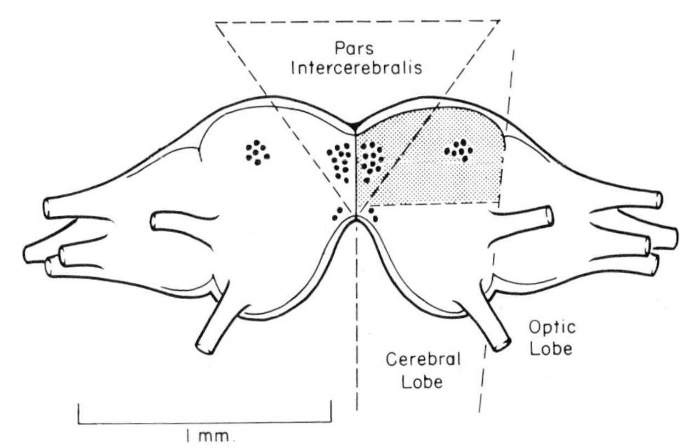

FIG. 2. A diagrammatic representation of the pupal brain of the Cecropia silkworm. The stippled area indicates the endocrinologically active region on each side. The solid circles show the location and number of the cell bodies of the medial (22), lateral (14), and posterior (4) neurosecretory cells as described by Herman and Gilbert (1965).

attention will be focused exclusively on the above-mentioned brain hormone, which is necessary for metamorphosis.

There is convincing evidence that this hormone is synthesized in the cell bodies of certain of the neurosecretory cells and that it escapes from the brain by traveling down the axons of these very same neurons. In most insects the axons emerge from the back of the brain as two pairs of short, tiny nerves. These nerves come together in a pair of discrete organs, the corpora cardiaca, and it is from these "neurohemal organs" that the hormone is normally released into the blood. If the corpora cardiaca are excised, the proximal ends of the neurosecretory axons often regenerate the neurohemal mechanism. The same appears to be true when loose brains are implanted into brainless pupae (Stumm-Zollinger, 1957).

Until ten years ago, the brain hormone resisted all efforts to obtain it apart from the living insect. Despite this handicap, the function of the hormone was worked out in considerable detail by biological maneuvers such as the surgical removal and transplantation of brains.

The chemistry of brain hormone remains a matter of controversy. After several false starts, there is now general agreement that it is a water-soluble, nondialyzable, heat-stable compound with a molecular weight in excess of 10,000. It is generally presumed to be a protein. However, this conclusion is by no means a certainty. There are substantial reasons for suspecting it to be a mucopolysaccharide or, at least, a mucoprotein (Williams, 1967a).

THE PROTHORACIC GLANDS: ECDYSONE

The primary function of the brain hormone is to activate another endocrine organ, the prothoracic glands (Williams, 1952, 1958). Under the tropic stimulation of brain hormone, the prothoracic glands secrete ecdysone.

When ecdysone is secreted and acts unopposed, it provokes a developmental response accompanied by the turning-on of many "new" genes and the turning-off of many "old" genes. The net effect is that the larval cells pupate and the pupal cells undergo adult differentiation.

What happens if the prothoracic glands stop secreting ecdysone? In that event, growth and metamorphosis come to an abrupt halt. Nature has exploited this state of affairs to provide for the overwintering of insects in the diapausing condition. After a long winter's nap, ecdysone is again secreted and development resumes where it had left off. Though the immediate cause of diapause is the lack of ecdysone, this deficiency can usually be traced back to the brain, and more particularly, the absence of sufficient brain hormone to "drive" the prothoracic glands.

Seventeen years ago, the German investigators Butenandt and Karlson launched a massive effort to extract and characterize ecdysone. They finally succeeded in isolating 25 mg of crystalline α-ecdysone from a ton of silkworms. They also obtained a trace of a more polar material, β-ecdysone (Butenandt and Karlson, 1954; Karlson, 1956). In assays carried out at Harvard University, the pure materials were found to do all things anticipated of the hormonal secretion of the prothoracic glands (Williams, 1954).

Eleven years elapsed before the chemistry of α-ecdysone was finally worked out by X-ray diffraction (Huber and Hoppe, 1965; Karlson *et al.*, 1965). To everyone's surprise, it proved to be the exotic sterol illustrated in Fig. 4. Its resemblance to cholesterol (Fig. 3) is self-evident.

Once the structure of ecdysone was known, chemists on three continents undertook the difficult task of synthesizing the molecule. An American group (Harrison *et al.*, 1966; Siddall *et al.*, 1966) and a team of German and Swiss scientists (Kerb *et al.*, 1966) simultaneously announced successful syntheses. Subsequently, a third syn-

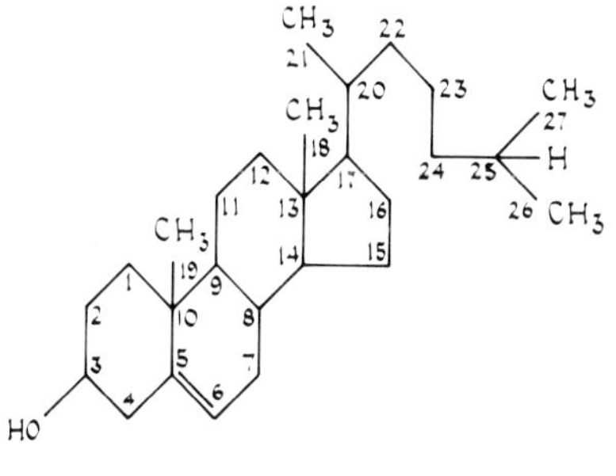

FIGS. 3 and 4. Structural formulas of α-ecdysone and its parental compound, cholesterol.

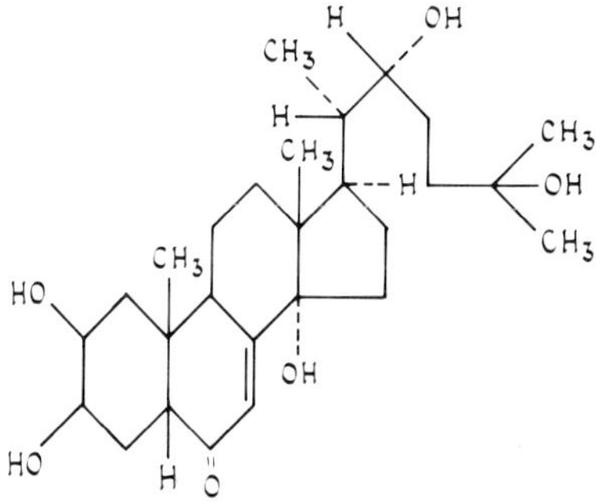

thesis was described by Japanese chemists (Mori *et al.*, 1968). Meanwhile, β-ecdysone was found to have the same structure as α-ecdysone except that an additional hydroxyl group is present at C-20 (Hampshire and Horn, 1966).

THE PHYTOECDYSONES

The synthesis of ecdysone proved to be so difficult and the yields so small that it seemed as though only vanishingly small amounts would ever be available for study. This discouraging prospect was changed overnight in a series of remarkable findings at the Tohoku University in Japan. There, Professors Nakanishi and Takemoto discovered that certain plants contain amazing amounts of ecdysone-like materials including authentic β-ecdysone (for summaries, see Nakanishi, 1968; Williams, 1970). These findings were soon confirmed and extended by several laboratories in Europe and the United States. It is now firmly established that certain plants have gone in for the synthesis and accumulation of impressive amounts of these complicated, hormonally active sterols.

As summarized elsewhere (Williams, 1970), ferns (Polypodiaceae) and certain families of gymnosperms (Taxodiaceae, Podocarpaceae) are particularly rich in phytoecdysones. Consider, for example, that Butenandt and Karlson obtained 25 mg of α-ecdysone from a ton of silkworms. This amount of β-ecdysone can be recovered from 25 gm of the dried leaves or roots of the yew tree (*Taxus baccata*) (Staal, 1967a,b) or from less than 2.5 gm of the dried rhizomes of the common fern, *Polypodium vulgare* (Jizba *et al.*, 1967). By latest count, 25 different phytoecdysones have been isolated and chemically characterized, and this number is certain to increase. With a few exceptions, all possess the same chemistry in the sterol ring system and differ among themselves in terms of substituent groups in the side chain.

Many of the phytoecdysones are superhormones in the sense of acting in lower concentrations than do α- or β-ecdysones. For example, cyasterone is twenty times as active as α-ecdysone when injected in diapausing silkworm pupae (Williams, 1968). This greater potency is due to the resistance of cyasterone to inactivation. Thus, in pupae of the Cynthia silkworm the time for 50% inactivation is 7 hours for α-ecdysone and 32 hours for cyasterone (Ohtaki and Williams, 1970).

PATHOLOGICAL EFFECTS OF EXCESSIVE ECDYSONE

When injected into diapausing silkworm pupae (*Samia cynthia*) in critically low doses, the phytoecdysones provoke the termination of dormancy and the formation of apparently normal moths. But, for all materials except α-ecdysone, a doubling of the critical dose

FIG. 5. Ventral view of an abnormal moth (*Samia cynthia*) formed from a diapausing pupa injected with 10 μg of the phytoecdysone cyasterone.

results in the formation of grossly abnormal and nonviable creatures such as illustrated in Fig. 5 (Kobayashi *et al.*, 1967a,b; Williams, 1968).

The pathological effects of excessive hormone are associated with a remarkable acceleration of development during the first few days after the injection. This fact is illustrated in Fig. 6, where the upper horizontal line records the sequence of events when normal development is triggered by the injection of a low dose of cyasterone (0.2 μg). The lower horizontal line illustrates the timing of these same events when extremely abnormal development is provoked by high doses of cyasterone (10 μg). Excessive hormone drives the early phase of metamorphosis so fast that, counting from the moment of injection, the sequence of events which normally requires 11 days is compressed into 4 or 5. Then, on about the third day of development, the precocious deposition of new cuticle locks all epidermal tissues in whatever stage they have managed to attain. As diagrammed in Fig. 6, the normal pace of development is resumed after about the third day. But the damage has already been done and the pupa will never form a viable moth. The phenomenon is of great interest and

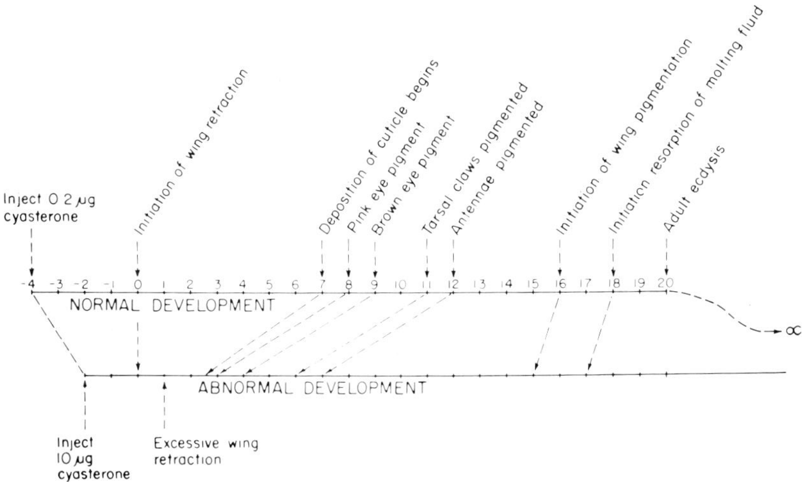

FIG. 6. The top line records the timetable of successive developmental events at 25°C when normal development of a Cynthia pupa is provoked by the injection of a physiological dose of ecdysone (in this case 0.2 μg of cyasterone). The lower horizontal line illustrates the timing of the same events when extremely abnormal development is provoked by the injection of an excessive dose of phytoecdysone (in this case 10 μg of cyasterone). After Williams (1968).

constitutes the first clear-cut case of hyperhormonism in any invertebrate.

MODE OF ACTION OF ECDYSONE

Ecdysone has come to be an extremely famous hormone because of its ability to provoke the formation of certain specific "puffs" in the giant chromosomes of larval Diptera (Clever and Karlson, 1960). This phenomenon (which has subsequently been confirmed in numerous laboratories) called attention to the possibility that hormones may have a clean-shot at the genome. It is fair to say that this prospect has transformed the complexion of endocrinology. As embodied in all updated texts, the hypothesis is that ecdysone gets into a cell, combines with certain genes, and turns on the new information prerequisite for metamorphosis.

The Swiss investigator Kroeger (1963) has advanced an alternative explanation which emphasizes the effects of ecdysone on cell permeabilities and, more particularly, the flux of sodium and potassium across cell membranes. According to Kroeger's theory, the formation of chromosomal puffs is a sequel to changes which ecdysone provokes in the "intracellular ionic milieu" (Kroeger and Lezzi, 1966).

Recent investigations carried out on a totally different system have confirmed the impressive effects of ecdysone on permeability (Kambysellis and Williams, 1969). In the studies in question, the testes of diapausing Cynthia pupae were subjected to *in vitro* culture in the presence or absence of ecdysone. The effects of the hormone were appraised in terms of the developmental response of the tens of thousands of primary spermatocytes which, as clusters of spherical "cysts," float around freely in the fluid that fills the testicular chambers.

Ecdysone was found to promote the transformation of spermatocytes into spermatozoa only when the culture medium also contained a much larger molecule which we have provisionally termed the "macromolecular factor" (MF). In cultures of intact testes, both ecdysone and MF proved to be necessary for the developmental response. The "payoff," long anticipated in the studies of Schmidt and Williams (1953), was the finding that MF was fully effective when the cysts were removed from the pupal testes and cultured "naked." Under that circumstance, the presence or absence of ecdysone was inconsequential. All that mattered was the presence of MF.

In additional experiments we found that intact testes, cultured for as little as 1 hour in the presence of ecdysone, showed the full developmental response when rinsed and cultured in ecdysone-free medium containing MF.

The picture that takes shape suggests a permissive role for ecdysone in terms of its ability to influence the permeability of the testes. It is obviously premature to generalize to other tissues and organs. However, we can state with assurance that the mode of action of ecdysone in the testes is in accord with several aspects of the theory proposed by Kroeger.

THE MACROMOLECULAR FACTOR

The biological assay for MF is its ability to provoke spermiogenesis in cultures of spermatocytal cysts obtained from the testes of diapausing male pupae of *S. cynthia*. If MF is absent, no development occurs. The presence of MF is signaled within 24 hours by the initiation of meiosis. After an additional 24 hours the cysts begin to elongate and to form bundles of spermatids and spermatozoa.

By the use of this test system, MF was found to be a heat-labile, nondialyzable substance which can be precipitated from "active blood" by the addition of 50 parts of water. The primary source of MF in the living insect is unknown. In diapausing pupae it seems to preexist in some latent or sequestered form since, in a matter of hours, it can be mobilized into the blood in response to integumentary injury. A case in point is its appearance in the blood of diapausing pupal abdomens that were isolated and stored for 24–48 hours at 25°C or at 2°C or even in the absence of oxygen. We suspect that MF is stored in the hemocytes and that it is released in response to several conditions: integumentary injury, or prolonged storage of pupae at 25°C, or the injection or secretion of ecdysone.

GROWTH WITHOUT METAMORPHOSIS: JUVENILE HORMONE AND THE STATUS QUO

Up to this point we have considered insect metamorphosis with special reference to its control by the nervous and endocrine systems. But the life story of insects is not one of ceaseless change and revolution. In point of fact, the strategy of insect development is to postpone metamorphosis until all necessary materials are at hand to construct a sexually competent adult. For example, the Cecropia silkworm grows for about 6 weeks prior to metamorphosis; during that period it molts its cuticle on four separate occasions and increases its initial mass some 5000-fold.

It is easy to show that ecdysone is the necessary stimulus for larval growth and molting. But, evidently, there is at work a conservative force which suppresses metamorphosis—that blocks "growing up" without interfering with growth in an unchanging state.

This conservative force proves to be "juvenile hormone"—an agent we have not had to mention up to this point. The hormone is synthesized by a pair of tiny cephalic glands, the corpora allata; these glands are also responsible for regulating its release into the blood. The action of juvenile hormone is to alter the cellular response to ecdysone—to suppress new synthetic acts without interfering with the use and re-use of genetic information already at the disposal of the cells and tissues. If this "brake" on progressive differentiation is removed by excising the corpora allata, the immature larva reacts to ecdysone by undergoing precocious metamorphosis to form a miniature pupa.

Under normal circumstances metamorphosis is postponed until late in larval life when the corpora allata, for some unexplained reason, lose their ability to secrete juvenile hormone. Then, for the first time, ecdysone can act in the presence of little or no juvenile hormone. The net effect is to turn off the "larval" genes and turn on the "pupal" genes. The "reading frame" racks forward to the next gene set (Williams, 1963a).

Particularly spectacular are the effects of juvenile hormone on the transformation of a pupa into an adult—a terminal phase of metamorphosis which proceeds in the presence of ecdysone and the absence of juvenile hormone. If juvenile hormone is supplied by the implantation of active corpora allata, the pupa is prevented from forming an adult. If an excess of hormone is caused to be present, the pupa molts into a second pupa (Fig. 7).

The phenomenon has its counterpart in the lower insects in which the mature larva transforms into a winged adult without traversing a pupal stage. Here again, the implantation of active corpora allata at the outset of the final larval stage is fully effective in suppressing adult differentiation.

THE CECROPIA JUVENILE HORMONE

For 20 years juvenile hormone remained a will-o'-the-wisp and resisted all efforts to extract or obtain it apart from the living insect. Then, about 13 years ago, a rich depot of the hormone was discovered in the abdomens of male Cecropia moths (Williams, 1956,

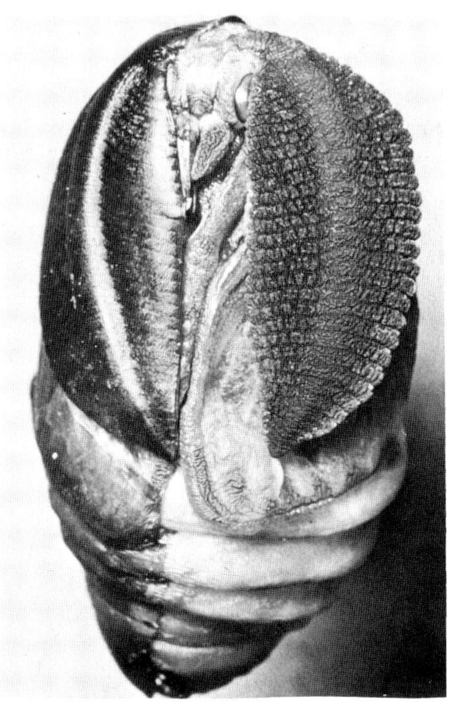

FIG. 7. Ventral view of a Polyphemus pupa caused to undergo a second pupal stage when juvenile hormone was supplied. The old pupal cuticle has been excised on one side and left *in situ* on the other side. After Williams (1959).

1963b). This was indeed a strange finding, and to this day only the closely related male Cynthia moth has been found to accumulate the hormone in this way. To extract the hormone, all one has to do is to excise the abdomens from male Cecropia or Cynthia moths, blend them in diethyl ether, then filter the solution, wash it in water, and evaporate the ether. One thereby obtains a golden oil, ca. 0.2 ml per abdomen, which shows impressive hormonal activity. The active principle in the golden oil has proved to be a heat-stable, water-insoluble, uncharged substance (Williams, 1956).

When injected into pupal silkworms, the crude extract duplicated all the effects previously realized when juvenile hormone was supplied by the implantation of active corpora allata. Indeed, it soon appeared that it was not necessary to inject the hormone: it sufficed merely to place the oily extract on the unbroken skin through which it promptly penetrated. The net result was the formation of nonviable creatures in which some cells had undergone metamorphosis and others had not.

This derangement of metamorphosis, coupled with the extract's activity on topical application, provided the first indication that juvenile hormone had potentialities as an insecticide (Williams, 1956, 1967b).

Until 1965 all attempts to isolate the pure hormone from Cecropia oil were unsuccessful (Williams and Law, 1965). Then, after three years of intensive effort, this difficult task was finally accomplished by a team of scientists headed by Herbert Röller (Röller and Bjerke, 1965) at the University of Wisconsin. Soon thereafter, its chemical structure was elucidated by mass spectrometry and other analytical studies performed on less than 300 μg of pure material (Röller *et al.*, 1967).

The empirical formula of Cecropia hormone is $C_{18}H_{30}O_3$, corresponding to a molecular weight of 294. It proves to be the methyl ester of the epoxide of a previously unknown fatty acid derivative (Fig. 8). The apparent simplicity of the molecule is deceptive. It has two double bonds and an oxirane ring and can therefore exist in eight geometric isomers plus two optical isomers, making a total of sixteen possible configurations. Moreover, with two ethyl groups attached to carbons 7 and 11, its synthesis from any known terpenoid is impossible.

The Wisconsin workers succeeded in synthesizing small amounts of the racemic mixture from which they separated and assayed the individual stereo isomers (Dahm *et al.*, 1967, 1968). In this manner they were able to show that the two double bonds in the authentic hormone are *trans,trans*, whereas the oxirane ring is *cis*. The optical isomerism has not been resolved.

According to Röller and Dahm (1968), the *trans* configuration of both double bonds is crucial for biological activity. By contrast, the stereochemistry of the oxirane ring seems to be of secondary importance. The authentic juvenile hormone, as well as the *dl* hormone, are active at submicrogram levels when assayed on most insects. Curiously enough, the synthetic ethyl ester proves to be eight times as active as the native methyl ester.

Subsequently, Meyer *et al.* (1968) isolated from Cecropia oil a second juvenile hormone which is identical in biological activity and structure to that described by Röller except that a methyl (rather than an ethyl) group is present at C-7 (Fig. 9). This variant of the hormone is responsible for 13–20% of the endocrine activity of Cecropia oil.

JUVENILE HORMONE AND EMBRYONIC DEVELOPMENT

The transformation of an insect egg into a first-stage larva is a metamorphosis every bit as impressive as anything occurring during postembryonic life. That being so, it is worth inquiring as to whether juvenile hormone can interfere with the sequential gene action prerequisite for embryonic development. Three years ago, Sláma and I (1966) showed that this is indeed the case. When we placed the eggs of the bug, *Pyrrhocoris apterus*, in contact with paper containing a juvenile hormone analog, embryonic development proceeded to a certain stage and then stopped with the still-living embryo short of completion. The most affected eggs contained a small disk of embryonic tissue afloat on a mass of unincorporated yolk. In other cases, the embryonic mass showed the rudiments of certain appendages but not of others. Sometimes parts of appendages had formed in the absence of the remainder. Needless to say, none of these eggs ever hatched.

Freshly laid eggs were most sensitive to inhibition by juvenile hormone. This finding suggested that the eggs might be yet more sensitive prior to oviposition. To test this possibility, juvenile hormone

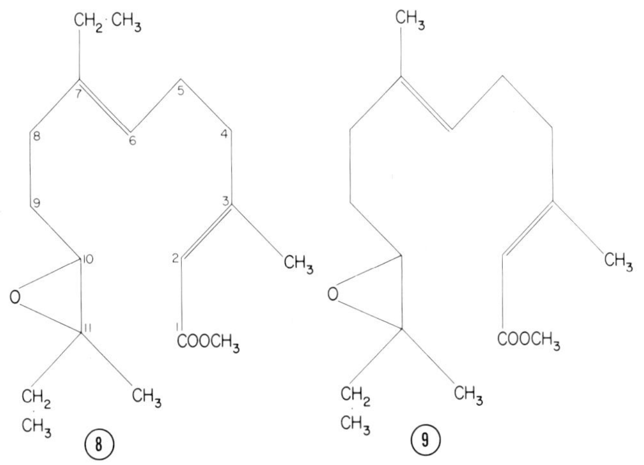

FIG. 8. Structural formula of juvenile hormone isolated by Röller *et al.* (1967) from male Cecropia moths.

FIG. 9. Structural formula of juvenile hormone isolated by Meyer *et al.* (1968) from male Cecropia moths.

was topically applied to a series of adult female *Pyrrhocoris*. Without exception, the eggs deposited by the treated animals behaved as though they had been directly exposed to the hormone; that is, development proceeded to a certain stage and then stopped. This implies that, after its topical application to adult females, the hormone is absorbed, translocated to the reproductive tract, and somehow gets into the eggs.

These simple experiments provided the first indication that the embryonic development of insects is subject to inhibition by the very same materials that can block metamorphosis. Subsequently, the ovicidal action of numerous juvenile hormone analogs has been confirmed for many other species of insects.

In no case is embryonic development blocked prior to the formation of the blastoderm—a stage in which the embryo consists of a yolk-filled blastula.

Riddiford and Williams (1967) called attention to the fact that unfertilized ova can be caused to undergo blastoderm formation by subjecting them to artificial parthenogenesis. Evidently, the unfertilized ovum has at its disposal the information for forming a blastoderm. Embryonic development beyond that early stage requires an activation of the embryo's own genome and a systematic derepression and "read-off" of specific genes, operons, and polyoperons.

In the embryo, as in the postembryonic insect, it is precisely this process which is blocked by the status quo action of juvenile hormone. Under both circumstances, the genetic information which is operational at the time of treatment remains fully operational. But in the presence of high titers of juvenile hormone, the insect is unable to activate the next gene set in its programmed life history.

REFERENCES

Butenandt, A., and Karlson, P. (1954). Über die Isolierung eines Metamorphose-Hormons der Insekten in kristallisierter Form. *Z. Naturforsch.* **9b,** 389–391.

Clever, U., and Karlson, P. (1960). Induktion von Puff-Veränderungen in den Speicheldrüsenchromosomen von *Chironomus tentans* durch Ecdyson. *Exptl. Cell. Res.* **20,** 623–626.

Dahm, K. H., Trost, B. M., and Röller, H. (1967). The juvenile hormone. V. Synthesis of the racemic juvenile hormone. *J. Am. Chem. Soc.* **89,** 5292–5294.

Dahm, K. H., Röller, H., and Trost, B. M. (1968). The juvenile hormone. IV. Stereochemistry of juvenile hormone and biological activity of some of its isomers and related compounds. *Life Sci.* **7,** 129–137.

Hampshire, F., and Horn, D. H. S. (1966). Structure of crustecdysone, a crustacean moulting hormone. *Chem. Comm.*, No. 2, 37–38.

Harrison, I. T., Siddall, J. B., and Fried, J. H. (1966). Steroids CCXCVII. Synthetic studies on insect hormones. Part III. An alternative synthesis of ecdysone and 22-isoecdysone. *Tetrahedron Letters* **29,** 3457–3460.

HERMAN, W. S., and GILBERT, L. I. (1965). Multiplicity of neurosecretory cell types and groups in the brain of the saturniid moth *Hyalophora cecropia* (L). *Nature* **205,** 926–927.

HUBER, R., and HOPPE, W. (1965). Zur Chemie des Ecdysons. VII. Die Kristall- und Molekülestrukturanalyse des Insektenverpuppungshormons Ecdyson mit der automatisierten Faltmolkülmethode. *Chem. Ber.* **98,** 2403–2424.

JIZBA, J., HEROUT, V., and SORM, F. (1967). Isolation of ecdysterone (crustecdysone) from *Polypodium vulgare* L. rhizomes. *Tetrahedron Letters* **18,** 1869–1891.

KAMBYSELLIS, M., and WILLIAMS, C. M. (1969). *In vitro* action of ecdysone. *Proc. Natl. Acad. Sci. U.S.* **63,** 231.

KARLSON, P. (1956). Biochemical studies on insect hormones. *Vitamins Hormones*, **14,** 227–266.

KARLSON, P., HOFFMEISTER, H., HUMMEL, H., HOCKS, P., and SPITELLER, G. (1965). Zur Chemie des Ecdysons. VI. Reaktionen des Ecdysonmoleküls. *Chem. Ber.* **98,** 2394–2402.

KERB, U., SCHULZ, G., HOCKS, P., WIECHERT, R., FURLENMEIER, A., FÜRST, A., LANGEMANN, A., and WALDVOGEL, G. (1966). Zur Synthese des Ecdysons. IV. Die Synthese des natürlichen Häutungshormons. *Helv. Chim. Acta* **49,** 1601–1606.

KOBAYASHI, M., TAKEMOTO, T., OGAWA, S., and NISHIMOTO, N. (1967a). The moulting hormone activity of ecdysterone and inokosterone isolated from *Achyranthis* radix. *J. Insect Physiol.* **13,** 1395–1399.

KOBAYASHI, M., NAKANISHI, K., and KORCEDA, M. (1967b). The moulting activity of ponasterones on *Musca domestica* (Diptera) and *Bombyx mori* (Lepidoptera). *Steroids* **9,** (5), 529–536.

KOPEČ, S. (1922). On the necessity of the brain for the inception of metamorphosis in insects. *Biol. Bull.* **42,** 323–342.

KROEGER, H. (1963). Chemical nature of the system controlling gene activities in insect cells. *Nature* **200,** 1234–1235.

KROEGER, H., and LEZZI, M. (1966). Regulation of gene action in insect development. *Ann. Rev. Entomol.* **11,** 1–22.

MEYER, A. S., SCHNEIDERMAN, H. A., HANZMANN, E., and KO, J. H. (1968). The two juvenile hormones from the Cecropia silk moth. *Proc. Natl. Acad. Sci. U.S.* **60,** 853–860.

MORI, H., SHIBATA, K., TSUNEDA, K., and SAWAI, M. (1968). Synthesis of ecdysone. *Chem. Pharm. Bull.* (*Tokyo*) **16,** 563–566.

NAKANISHI, K. (1968). Conference on insect-plant interactions. *BioScience* **18,** 791–799.

NÜESCH, H. (1952). Ueber den Einfluss der Nerven auf die Muskelentwichlung bei *Telea polyphemus*. *Rev. Suisse Zool.* **59,** 294–301.

NUESCH, H. (1968). The role of the nervous system in insect morphogenesis and regeneration. *Ann. Rev. Entomol.* **13,** 27–44.

OHTAKI, T., and WILLIAMS, C. M. (1970). Inactivation of ecdysone and phytoecdysones. *In preparation.*

RIDDIFORD, L. M., and WILLIAMS, C. M. (1967). The effects of juvenile hormone analogues on the embryonic development of silkworms. *Proc. Natl. Acad. Sci. U.S.* **57,** 595–601.

RÖLLER, H., and BJERKE, J. S. (1965). Purification and isolation of juvenile hormone and its action in lepidopteran larvae. *Life Sci.* **4,** 1617–1624.

RÖLLER, H., and DAHM, K. H. (1968). The chemistry and biology of juvenile hormone. *Recent Progr. Hormone Res.* **24,** 651–680.

RÖLLER, H., DAHM, K. H., SWEELY, C. C. and TROST, B. M. (1967). The structure of the juvenile hormone. *Angew. Chem.*, Intern. Ed. **6,** 179–180.

SCHMIDT, E. S., and WILLIAMS, C. M. (1953). Physiology of insect diapause. V. Assay of the growth and differentiation hormone of Lepidoptera by the method of tissue culture. *Biol. Bull.* **105,** 174–187.

SIDDALL, J. B., CROSS, A. D., and FRIED, J. H. (1966). Steroids. CCXCII. Synthetic studies on insect hormones. II. The synthesis of ecdysone. *J. Am. Chem. Soc.* **88,** 862–863.

SLÁMA, K., and WILLIAMS, C. M. (1966). "Paper factor" as an inhibitor of the embryonic development of the European bug, *Pyrrhocoris apterus*. *Nature* **210,** 329–330.

STAAL, G. B. (1967a). Plants as a source of insect hormones. *Koninkl. Ned. Akad. Wetenschap. Proc. Ser. C,* **70,** 409–418.

STAAL, G. B. (1967b). Insect hormones in plants. *Mededel. Rijksfacult. Landbouwwetenschap. Gent* **32,** 393–400.

STUMM-ZOLLINGER, E. (1957). Histological study of regenerative processes after transection of the nervi corporis cardiaci in transplanted brains of the Cecropia silkworm (*Platysamia cecropia* L). *J. Exptl. Zool.* **134,** 315–326.

WIGGLESWORTH, V. B. (1940). The determination of characters at metamorphosis in *Rhodnius prolixus* (Hemiptera). *J. Exptl. Biol.* **17,** 201–222.

WILLIAMS, C. M. (1948). Extrinsic control of morphogenesis as illustrated in the metamorphosis of insects. *Growth Symp.* **12,** 61–74.

WILLIAMS, C. M. (1952). Physiology of insect diapause. IV. The brain and prothoracic glands as an endocrine system in the Cecropia silkworm. *Biol. Bull.* **103,** 120–138.

WILLIAMS, C. M. (1954). Isolation and identification of the prothoracic gland hormone of insects. *Anat. Record* **120,** 743.

WILLIAMS, C. M. (1956). The juvenile hormone of insects. *Nature* **178,** 212–213.

WILLIAMS, C. M. (1958). Hormonal regulation in insect metamorphosis. *Symp. Chem. Basis of Develop., Baltimore, 1958. John Hopkins Univ. McCollum-Pratt Inst. Contrib.* **234,** 794–806.

WILLIAMS, C. M. (1963a). Differentiation and morphogenesis in insects. *In* "The Nature of Biological Diversity" (J. M. Allen, ed.), pp. 243–260. McGraw-Hill, New York.

WILLIAMS, C. M. (1963b). The juvenile hormone. III. Its accumulation and storage in the abdomen of certain male moths. *Biol. Bull.* **124,** 355–367.

WILLIAMS, C. M. (1967a). The present status of the brain hormone. *In* "Insects and Physiology" (J. W. L. Beament and J. E. Treherne, eds.), pp. 133–142. Oliver & Boyd, London.

WILLIAMS, C. M. (1967b). Third-generation pesticides. *Sci. Am.* **217,** 13–17.

WILLIAMS, C. M. (1968). Ecdysone and ecdysone-analogues: their assay and action on diapausing pupae of the Cynthia silkworm. *Biol. Bull.* **134,** 344–355.

WILLIAMS, C. M. (1970). Hormonal interactions between plants and insects. *In* "Perspectives in Chemical Ecology" (E. Sondheimer, ed.), Academic Press, New York. (in press).

WILLIAMS, C. M., and LAW, J. H. (1965). The juvenile hormone. IV. Its extraction, assay, and purification. *J. Insect Physiol.* **11,** 569–580.

WILLIAMS. C. M., and SCHNEIDERMAN, H. A. (1952). The necessity of motor innervation for the development of insect muscles. *Anat. Record* **113** (4), 54–55.

DEVELOPMENTAL BIOLOGY SUPPLEMENT 3, 151–171 (1969)

Estrogen–Receptor Interaction in Target Tissues

E. V. JENSEN, M. NUMATA, S. SMITH, T. SUZUKI, P. I. BRECHER, AND E. R. DESOMBRE

The Ben May Laboratory for Cancer Research, The University of Chicago, Chicago, Illinois 60637

INTRODUCTION AND BACKGROUND

It has long been recognized that certain mammalian tissues, associated with the reproductive process, are unable to grow and function optimally in the absence of minute amounts of steroidal compounds called sex hormones. During the course of differentiation, these hormone-dependent tissues appear either to have lost some vital capacity, which is restored to them by the steroid, or to have acquired some mechanism of growth restraint, which is relieved by the action of the hormone. Although the actual basis of hormone dependency remains obscure, an important difference between dependent and nondependent tissues was shown by studies of the fate in various tissues of tritiated estrogens administered to hormone-deprived animals. The "target" tissues, such as uterus, vagina, and anterior pituitary, were found to possess a striking affinity for estradiol, indicating that they contained specific estrogen-binding componenets called "estrogen receptors" or "estrophiles."

Detailed investigations[1] of the incorporation and retention of radioactive steroid, following the administration of physiological amounts of various estrogens, have established the principal characteristics of the estrogen–receptor interaction *in vivo* and have provided a basis for evaluating the significance of estrogen binding phenomena with *in vitro* tissue systems or with cell fractions. In the rat, the target tissues show strong affinity for estradiol, 17α-methylestradiol, 17α-ethynylestradiol, and hexestrol and a transient affinity for estriol; each of these substances appears to initiate uterine growth without undergoing chemical alteration, even though they undergo

[1] References to the earlier papers describing the characteristics of the binding of estrogenic hormones by target tissues, both *in vivo* and *in vitro*, are given by Hechter and Halkerston (1964) and Jensen *et al.* (1966, 1967).

extensive metabolic transformation elsewhere in the animal. In contrast, neither estrone nor mestranol bind to target tissues; the estrogenic action of these two substances appears to result from their conversion in the organism to estradiol and 17α-ethynylestradiol, respectively. Specific incorporation of estrogens into hormone-dependent tissues is blocked by certain antiuterotropic substances, such as ethamoxytriphetol (MER-25), nafoxidine (Upjohn-11,100), Parke-Davis CI-628, and clomiphene. Using various doses of nafoxidine, the inhibition of uterine growth was shown to parallel the reduction in estradiol binding, providing evidence for the importance of the estrogen–receptor association in the physiological action of the hormone. In contrast, actinomycin D and puromycin, both of which block overall growth as well as many early biochemical responses to estrogen, do not interfere with the binding of estradiol to target tissues, indicating that the estradiol–receptor interaction is an early step in the uterotropic process, preceding the acceleration of biosynthetic reactions sensitive to actinomycin D or puromycin.

Studies of estradiol incorporation as a function of dose indicate that the specific interaction of estrogens with target tissues in the immature rat consists of two distinct phenomena: an uptake process, which is not saturable even with hyperphysiological amounts of administered hormone, and a retention process, which becomes saturated as the dose exceeds the physiological level.

When surviving uterine horns are exposed to dilute (10^{-10} *M*) solutions of tritiated estradiol in a Krebs-Ringer buffer at 37°C, there is an association of hormone with receptor which shows the characteristics described above for the phenomenon *in vivo*. The simple *in vitro* system affords a useful method for studying estrogen–receptor interaction, uncomplicated by circulating estrogen metabolites and other plasma components. Using the *in vitro* system, it was shown that sulfhydryl groups play an important role in the estrogen–receptor interaction; treatment of uterine tissue with sulfhydryl-blocking reagents before exposure to estradiol prevents the specific hormone uptake whereas such treatment of uteri previously exposed to estradiol, either *in vivo* or *in vitro*, causes rapid release of the bound hormone. Sensitivity to sulfhydryl reagents or to specific inhibitors, such as nafoxidine or Parke-Davis CI-628, provides valuable criteria for distinguishing between the physiological estrogen-receptor interaction and artifacts of nonspecific estrogen binding to macromolecules in broken cell systems.

INTRACELLULAR LOCALIZATION OF ESTROGEN IN UTERUS

Cellular fractionation experiments, confirmed by autoradiographic investigations, show the existence of two sites of estrogen binding in target tissues (Fig. 1). When uterine homogenates from estradiol-treated rats are subjected to differential centrifugation, either in sucrose or in hypotonic medium, the steroid is found in two fractions (Talwar *et al.*, 1964; Noteboom and Gorski, 1965; Jensen, 1965; King *et al.*, 1965; King and Gordon, 1966; Baulieu *et al.*, 1967). Between 15 minutes and 6 hours after injection of radioactive hormone, a minor portion (20–25%) of the uterine estradiol is bound to a macromolecule appearing in the supernatant or cytosol fraction; most of the steroid is localized in the heavy or nuclear–myofibrillar fraction, from which it can be extracted as a soluble macromolecular complex by 0.3 *M* potassium chloride at pH 7.4 (Jungblut *et al.*, 1967; Jensen *et al.*, 1967) or, more effectively, by 0.4 *M* KCl at pH 8.5 (Puca and Bresciani, 1968).

Autoradiographic localization of tritiated estradiol in rat uterus, using a dry-mount technique which minimizes steroid translocation during tissue processing (Stumpf and Roth, 1966), demonstrates a distribution pattern similar to that obtained by cellular fractionation (Stumpf, 1968; Jensen *et al.*, 1967, 1969b). In all regions of the uterus most of the radioactivity is seen in the nuclei; the extranuclear radioactivity varies from 15–20% of the total in the epithelial glands and myometrium to 35–40% in the lamina propria, strengthening the as-

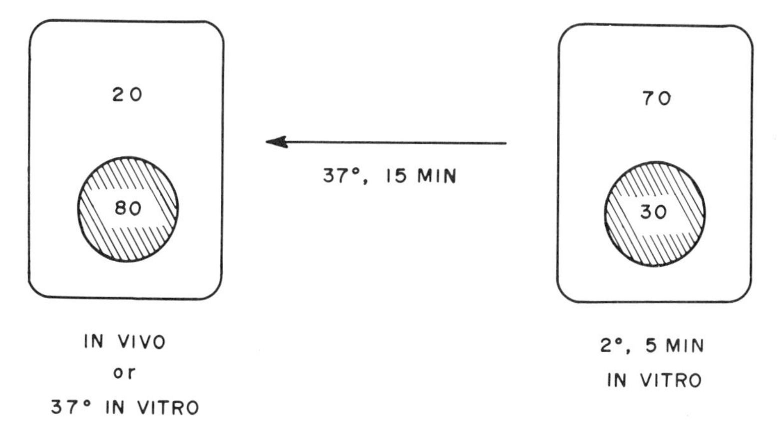

FIG. 1. Distribution of nuclear and cytosol radioactivity in rat uteri exposed to tritiated estradiol under various conditions.

sumption that the radioactivity of the supernatant fraction represents extranuclear estradiol. The correlation of the results of fractionation with those of autoradiography provides reassurance that the high estradiol content of the nuclear fraction is not an artifact of redistribution taking place during cell disruption.

After immature rat uteri have been incubated with tritiated estradiol at 37°C *in vitro*, the intracellular distribution pattern of uterine radioactivity, as determined either by fractionation or by autoradiography, is similar to that observed after hormone administration *in vivo* (Jensen *et al.*, 1967, 1969b). But if the tissue is exposed to hormone at 2°C, most of the radioactive steroid is present in the cytosol fraction and is seen in the extranuclear region on autoradiography. When such uteri are removed from the hormone solution and maintained in a moist atmosphere at 2°, there is little change in the radioactivity pattern. On warming to 37°C, rapid redistribution of the hormone takes place within uterine cells (Fig. 1) to yield a predominantly nuclear localization observed both by fractionation and by autoradiography (Jensen *et al.*, 1968, 1969a, b). The preponderance of cytosol radioactivity after exposure at 2°C *in vitro* has also been observed by Rochefort and Baulieu (1969) as well as by Gorski *et al.* (1968), who have described a similar shift of hormone from cytosol to nucleus on warming to 37°C. These observations indicate that, at least *in vitro*, radioactive steroid can bind to the cytosol in the cold and can pass from the cytosol to its nuclear binding site by a process which is temperature-dependent.

ESTROPHILIC SUBSTANCES OF UTERINE TISSUE

With the application by Toft and Gorski (1966) of sucrose density gradient ultracentrifugation to the study of bound estradiol in homogenate fractions, a valuable technique became available for detecting the estrophilic substances of target tissues and for determining their nature. These investigators found that the radioactivity of the cytosol fraction sediments as a discrete band (Fig. 2) with a coefficient, originally believed to be 9.5 S, but later observed by us and by others (Erdos, 1968; Rochefort and Baulieu, 1968) to be close to that of yeast alcohol dehydrogenase, or about 8 S. The 8 S protein shows a marked tendency to form large aggregates, slowly on standing and rapidly on attempted purification by ammonium sulfate precipitation or Sephadex G-200 filtration (Jensen *et al.*, 1967; Erdos, 1968; DeSombre *et al.*, 1969). This aggregation appears to involve in-

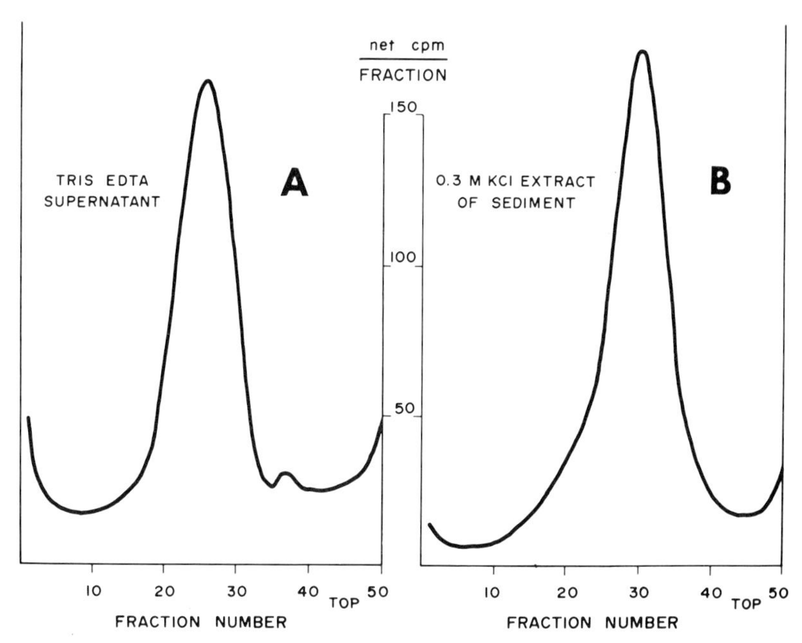

FIG. 2. Sedimentation patterns of two radioactive estradiol–receptor complexes from immature Sprague-Dawley rat uteri: (A) supernatant fraction; (B) nuclear extract. Fifty uterine horns (573 mg) excised 2 hours after subcutaneous injection of 0.045 μg (9.36 μCi) estradiol-6,7-^{3}H (E-2*) in 0.2 ml of saline to 24-day-old rats, were homogenized at 2°C (five 15-second glass pestle homogenization periods with 45 seconds intermittent cooling) in 2.3 ml of Tris-EDTA buffer (10 m*M* Tris plus 1.5 m*M* ethylenediaminetetraacetic acid), pH 7.4. The homogenate was centrifuged at 204,000 *g* for 1 hour at 2°C, the supernatant fraction was removed, and the sediment was rehomogenized gently at 2°C in 2.0 ml 0.3 *M* KCl in Tris-EDTA, pH 7.4, and centrifuged at 204,000 *g* for 1 hour to yield the nuclear extract. A 200-μl portion of either supernatant fraction or nuclear extract was layered on 4.8 ml of a 5–20% linear sucrose gradient containing Tris-EDTA, pH 7.4 (and, with nuclear extract, 0.3 *M* KCl). After centrifugation at 216,400 *g* (7-hour supernatant fraction; 10-hour nuclear extract), fifty 100-μl fractions were drawn off from the bottom for direct counting in dioxane–xylene scintillation mixture. Reproduced with permission from "Autoradiography of Diffusible Substances" (L. J. Roth and W. E. Stumpf, eds.), Academic Press, New York, 1969.

teraction with other components of the cytosol, since, after purification by preparative density gradient centrifugation, the 8 S complex is resistant to aggregation.

When the salt extract of uterine nuclei, which represents the major portion of the uterine estradiol, is subjected to similar sucrose density ultracentrifugation,[2] the radioactivity is found to sediment more slowly than that of the cytosol (Figs. 2 and 3) giving a sharp peak at about 5 S (Jungblut *et al.*, 1967; Jensen *et al.*, 1967). Exposure of uterine tissue to estradiol solutions *in vitro* gives rise to the same 8 S and 5 S estradiol–receptor complexes obtained when the hormone is given *in vivo*. The high level of cytosol radioactivity observed when estradiol uptake is effected at 2° *in vitro* (Fig. 1) sediments entirely as 8 S complex; after these tissues are warmed to 37°, the radioactivity which shifts to the nucleus is extractable as 5 S complex (Jensen *et al.*, 1968).

Both the 8 S and the 5 S receptor substances appear to be sulfhydryl-containing proteins, inasmuch as both complexes are destroyed by treatment with trypsin, pronase, or *p*-hydroxymercuribenzoate, but not by ribonuclease or deoxyribonuclease (Toft and Gorski, 1966; Jensen *et al.*, 1967, 1968). But there are important differences between the two complexes. When exposed to 0.3 or 0.4 *M* potassium chloride, as used to extract the 5 S complex from the nucleus, the 8 S receptor protein, with or without estradiol attached, is reversibly dissociated into a subunit (Erdos, 1968; Rochefort and Baulieu, 1968; Korenman and Rao, 1968; Jensen *et al.*, 1969a). Despite reports to the contrary, this subunit of the 8 S protein can be readily distinguished from the 5 S complex of the nuclear extract by high resolution gradient centrifugation in the presence of salt (Figs. 3 and 4). The nuclear complex sediments faster (5 S) and the cytosol subunit more slowly (4 S) than bovine plasma albumin (4.6 S), so that each complex can be recognized in the absence of the other by its relation to the albumin marker. In the absence of salt, the 4 S subunit reverts either to the 8 S form or to a higher aggregate, depending on the care with which salt is removed.

[2] In our earlier studies, as well as in experiments of certain other investigators (e.g., Puca and Bresciani, 1968), the nuclear radioactivity sedimented cleanly at about 5 S, whether or not KCl was present in the sucrose gradient. More recently we, as well as Korenman and Rao (1968), have found that the nuclear extract, presumably prepared in the same manner, can undergo aggregation unless salt is present in the gradient. The basis of this variable behavior is not clear.

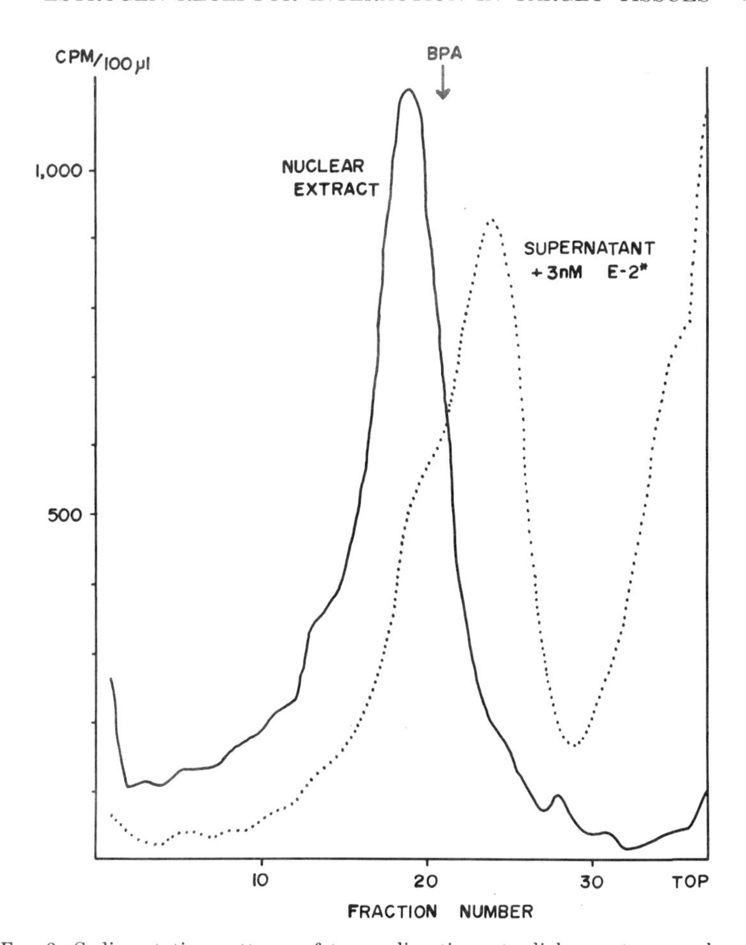

FIG. 3. Sedimentation patterns of two radioactive estradiol–receptor complexes in sucrose gradients containing KCl. Slit uterine horns from 25-day-old rats were stirred in KRH buffer, pH 7.3, for 15 minutes at 2° to remove serum proteins, homogenized in 4 volumes of Tris-EDTA buffer, pH 7.4, and centrifuged 30 minutes at 308,000 *g*. The supernatant fraction was diluted with one-half volume of Tris-EDTA buffer containing E-2* to give a final concentration of 3 n*M*. Nuclear extract was prepared by similarly homogenizing uteri, excised 2 hours after subcutaneous injection of 0.10 μg (20.7 μCi) E-2* in 0.2 ml saline, and extracting the sediment with 0.4 *M* KCl in Tris-EDTA, pH 7.4. A 200-μl aliquot of either the supernatant fraction (with added E-2*), nuclear extract or 0.5% bovine plasma albumin (BPA) containing 6 n*M* E-2* was layered on 3.6 ml of a linear 5–20% sucrose gradient containing 0.4 *M* KCl and Tris-EDTA, pH 7.4. After centrifugation for 10 hours at 308,000 *g*, 100-μl fractions were drawn off from the bottom for counting in a toluene–Triton X-100 scintillation mixture.

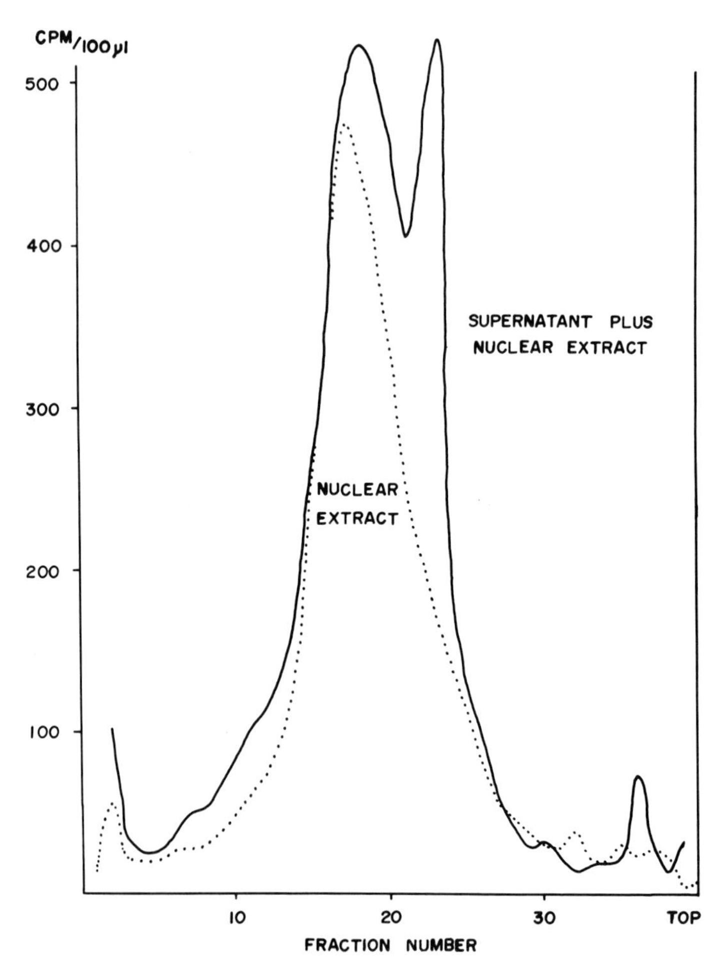

FIG. 4. Separation of mixture of nuclear and cytosol complexes on salt-containing sucrose gradient. Uterine horns, excised 1 hour after subcutaneous injection of 0.10 μg E-2* in 0.2 ml saline to 25-day-old rats, were stirred in KRH buffer, pH 7.3, for 15 minutes at 2°C and homogenized in 4 volumes of TKE buffer (10 mM Tris, 10 m*M* KCl and 1 m*M* EDTA) pH 7.4. The supernatant fraction and nuclear extract were prepared as described for Fig. 3. The nuclear extract was mixed with an equal volume of either supernatant fraction or extraction medium; 200 μl of each mixture was layered on a 5–20% sucrose gradient containing 0.4 *M* KCl and Tris-EDTA, pH 7.4, and centrifuged 14 hours at 308,000 *g*. Total counts per minute applied to gradient: nuclear extract, 5140; supernatant fraction, 2340.

If calcium ions are added to the dissociated cytosol complex, the 4 S subunit does not revert to the 8 S form when salt is removed (Fig. 5) and does not aggregate during ammonium sulfate precipitation, gel filtration, and column chromatography. Using these techniques, the calcium-stabilized 4 S unit of the estradiol–receptor complex from calf uterine cytosol has been purified several thousandfold to yield a product of about 2% purity, showing a molecular weight (Sephadex G-200 elution) of about 75,000 and an isoelectric point of 6.4 as compared to values of 200,000 and 5.8 obtained with the original 8 S complex (DeSombre *et al.*, 1969).

Although neither the 8 S nor the 5 S estradiol–receptor complex is destroyed by ribonuclease, the 8 S complex appears to associate with this enzyme to form a new entity, which sediments more rapidly and is resistant to the slow aggregation which the crude 8 S complex undergoes on incubation at 10°C (Fig. 6). When exposed to potassium chloride, this new entity is dissociated into a subunit which is indistinguishable by sedimentation velocity from the 4 S subunit of the original 8 S complex, suggesting that the binding to RNase is disrupted by the salt. Although it is doubtful that an interaction with RNase is of physiological significance in the mechanism of estrogen action, the ability of the 8 S complex to bind to basic proteins may represent a property important to its role in the growth initiation process.

A further difference between the two estrophilic proteins of uterine tissue lies in the fact that the 8 S complex of the cytosol can be formed directly, event at 2°C, simply by adding tritiated estradiol to the supernatant fraction of uteri not previously exposed to hormone (Toft *et al.*, 1967; Jungblut *et al.*, 1967; Jensen *et al.*, 1967, 1968).

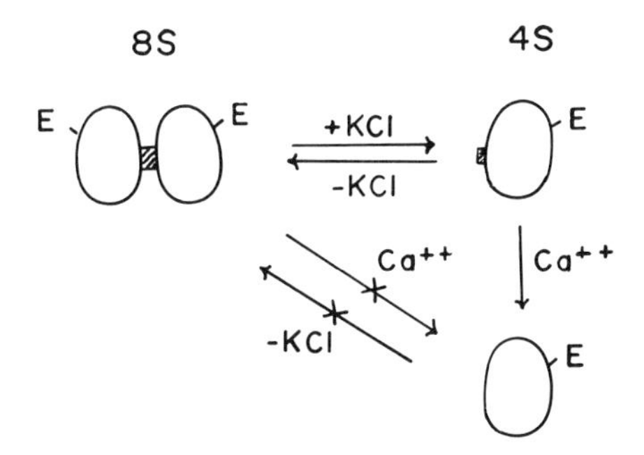

FIG. 5. Salt-induced deaggregation of cytosol receptor.

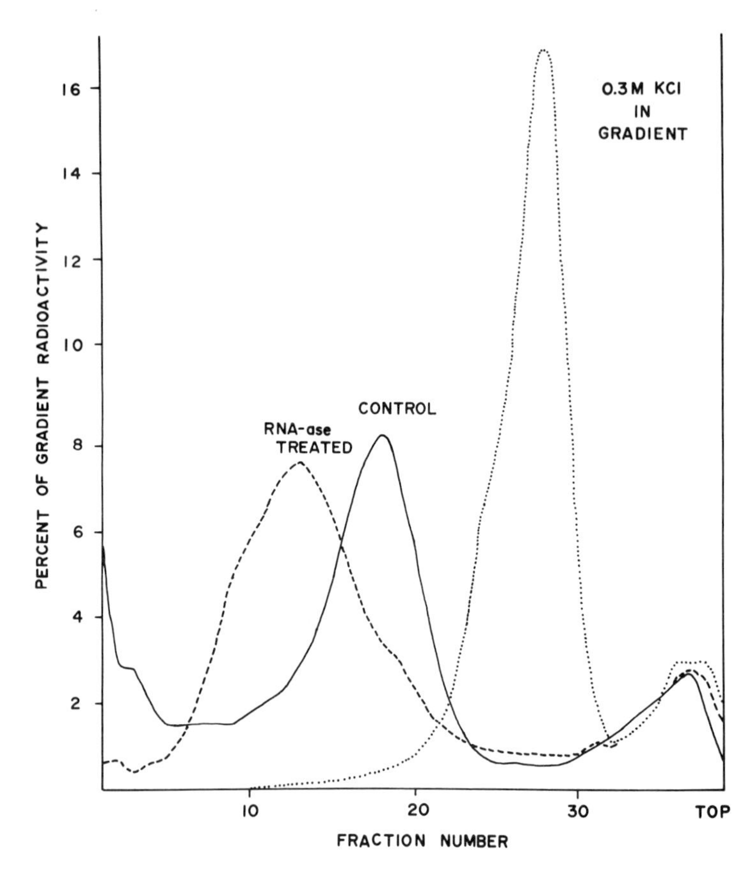

FIG. 6. Influence of ribonuclease on sedimentation pattern of uterine cytosol receptor. Uterine horns from 28-day-old rats were stirred in KRH buffer for 15 minutes at 2°C and homogenized in four volumes of Tris-EDTA buffer, pH 7.4. A 1.6-ml portion of the supernatant fraction was mixed with 0.4 ml E-2* in buffer (final concentration 4 n*M*), and aliquot portions (400 μl) were treated with 100 μl of buffer with and without 500 μg of ribonuclease (Worthington). After incubation for 30 minutes at 10°C, 200-μl aliquots of each mixture were layered on 5–20% sucrose gradients containing Tris-EDTA, pH 7.4, with and without 0.3 *M* KCl, and centrifuged 7.5 hours at 308,000 *g*. In the presence of KCl, the sedimentation patterns of RNase treated and non-treated samples were practically identical. Total radioactivity in each gradient varied from 20,840, to 21,880 cpm.

This phenomenon provides a convenient method for estimating the total 8 S receptor capacity of a supernatant fraction simply by determining the area of the 8 S sedimentation peak when sufficient estradiol has been added to saturate the receptor (Fig. 7). In this way it can be demonstrated that many rat tissues possess small amounts of the 8 S estrophile. The target tissues are unique in that they contain much higher levels of the binding protein than the nontarget tissues (Table 1).

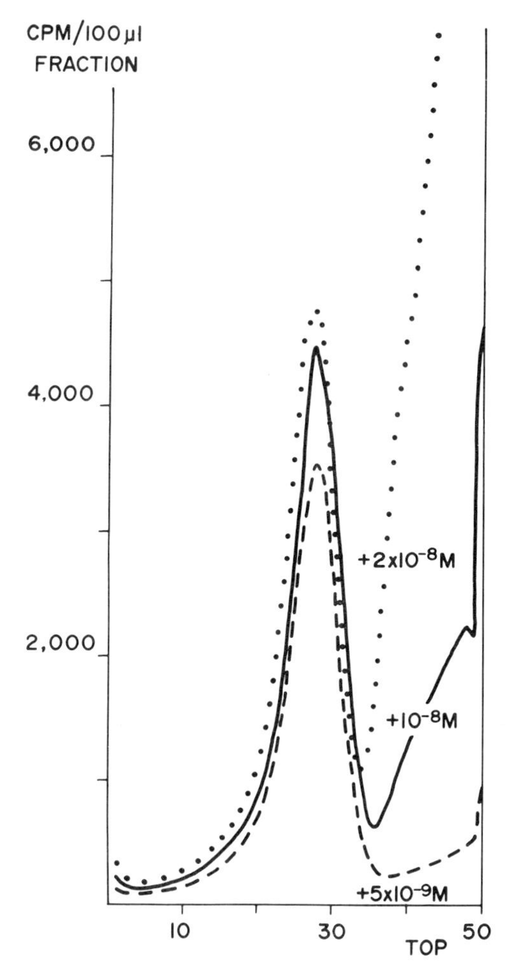

FIG. 7. Saturation of the 8S receptor protein. Fifty slit uterine horns (537 mg) from 23-day-old rats were stirred in KRH buffer, pH 7.3, for 30 minutes at 37°C and homogenized in 2.15 ml of cold Tris-EDTA buffer, pH 7.4. Aliquot portions (400 μl) of the 205,000 *g* supernatant were treated with 100-μl of buffer containing E-2* to give final concentrations of 5, 10, and 20 n*M*, respectively. A 200-μl portion of each mixture was layered on a 5–20% sucrose gradient containing Tris-EDTA buffer, pH 7.4 and centrifuged 7 hours at 216,000 *g*.

TABLE 1
8 S RECEPTOR CAPACITIES OF IMMATURE RAT TISSUES

Tissue	Percent of Uterine capacity[a]
Uterus	100
Vagina	50
Pituitary[b] (♀ or ♂)	30
Seminal vesicle	3
Testis	2
Ovary	2
Adrenal	2
Diaphragm	0.6
Muscle	0.5
Liver	0.4
Prostate	0.4
Kidney	0.3

[a] The 8 S receptor capacity of uterine cytosol can range from 30 to 100 femtomoles per milligram of wet tissue depending on the speed and method of preparation.

[b] Whole pituitary. The receptors are localized in the anterior lobe where their concentration is about the same as in vagina.

In contrast to 8 S complex of the cytosol, the 5 S complex of the nucleus is not produced by addition of estradiol either to the nuclear extract or to the nuclei themselves, although it can be formed in the whole homogenate (Jensen *et al.*, 1967). This observation first suggested that the 5 S estradiol–receptor complex of the nucleus was in some way derived from the 8 S complex of the cytosol and led to the formulation of the two-step interaction mechanism discussed in the following section.

TWO-STEP INTERACTION OF ESTRADIOL IN UTERUS

The fact that the interaction of estradiol with uterine tissue is a more complex phenomenon than mere binding of the hormone to a cellular receptor site became evident from a variety of experimental observations, some of which already have been mentioned. The recognition of nonsaturable uptake as distinct from saturable retention, the two intracellular sites of hormone localization yielding different estradiol–protein complexes, the direct formation of the 8 S complex in the cytosol but not the 5 S complex in the nuclear extract, and the fact that the 8 S receptor capacity is not saturated by physiological doses of estradiol, all suggested that the 8 S protein might function as an extranuclear "uptake receptor," bringing the hormone to the nucleus and playing some role in its fixation there (Jensen *et al.*, 1967).

Subsequent evidence has confirmed this concept and established a firm experimental basis for a two-step interaction of estradiol with uterine tissue. As discussed earlier (Fig. 1), the redistribution of radioactivity when uteri, originally exposed to estradiol *in vitro* at 2°C, are warmed to 37°C indicates that transfer of estradiol from the 8 S complex of the cytosol to the 5 S complex of the nucleus can take place and that this transformation is strongly temperature-dependent (Gorski *et al.*, 1968; Jensen *et al.*, 1968; Shyamala and Gorski, 1969). The presence of unused uptake receptor is evident from the large increase in the 8 S sedimentation peak when additional estradiol is added to uterine cytosol from animals which have received various doses of the hormone (Jensen *et al.*, 1968). For example, 15 minutes after injection of 0.05 μg of estradiol, only 5% of the 8 S protein in the cytosol is associated with steroid; even with a hyperphysiological dose (0.5 μg), less than half the binding capacity is utilized. Similar lack of saturation of 8 S receptors is observed 2 hours after injecting 0.05 or 0.2 μg of estradiol *in vivo* or after uptake of physiological or hyperphysiological amounts of hormone *in vitro*. This reserve capacity of 8 S estrophile is consistent with its postulated role in the nonsaturable uptake process.

In the foregoing experiments it was noted that the total estradiol binding capacity of the uterine supernatant fraction is less after a large dose of hormone than after a small one, suggesting that interaction with estradiol depletes the 8 S receptor content of the cytosol (Jensen *et al.*, 1968). This conclusion is substantiated by the rapid and progressive drop in total 8 S receptor content which takes place during the first 4 hours after the injection of a physiological dose of estrogen (Fig. 8). The subsequent rise in binding capacity probably reflects the synthesis of new receptor protein, inasmuch as it does not occur if cycloheximide is administered to the animal (Jensen *et al.*, 1969). With a massive dose of estradiol (5 μg) 90% of the 8 S protein disappears within 15 minutes. Similar depletion of 8 S receptor during exposure of uteri to estradiol *in vitro* has been observed both by us and by Gorski *et al.* (1968).

Although estradiol will bind to isolated uterine nuclei as well as to nuclei from other tissues, this interaction does not give rise to any extractable 5 S complex. Nor is the complex formed by adding estradiol to the extract of nuclei from uteri not previously exposed to

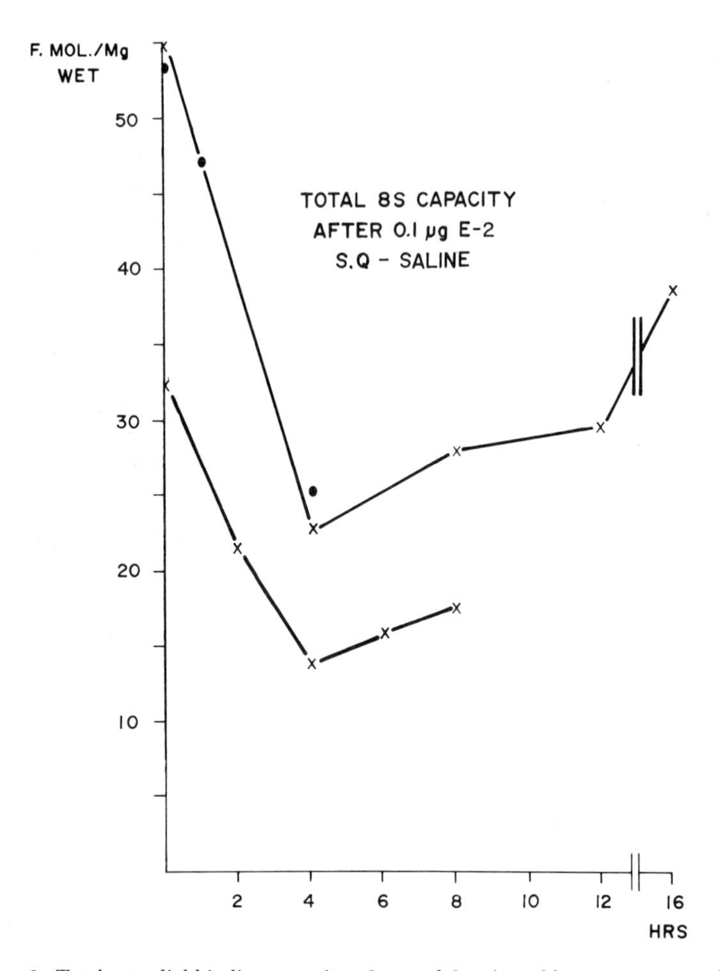

FIG. 8. Total estradiol-binding capacity of cytosol fraction of immature rat uteri excised at different times after subcutaneous injection of 0.10 μg E-2* in saline. Uteri (8 to 10 rats per group) were homogenized and supernatant fraction was prepared as described for Fig. 2, using 4 volumes of Tris-EDTA for experiment represented by solid circles and 9 volumes for two experiments represented by crosses. After addition of excess E-2* (10–20 n*M*), a 200-μl aliquot portion of each supernatant fraction was layered on a 5–20% sucrose gradient containing Tris-EDTA, pH 7.4, and centrifuged for 6.5 hours at 279,700 *g* or 8.5 hours at 204,000 *g*. The total radioactivity in the 8 S peak, plus a small amount of aggregated material, was measured and expressed as femtomoles of bound estradiol per milligram of original uterine tissue. Reproduced by permission from "Autoradiography of Diffusible Substances" (L. J. Roth and W. E. Stumpf, eds.), Academic Press, New York, 1969.

estradiol. But if uterine nuclei or lysed nuclear sediment (from homogenization in hypotonic buffer) are incubated with estradiol in the presence of the cytosol fraction, radioactivity taken up by the nuclei is extractable as 5 S complex (Jensen *et al.*, 1968, 1969a). The formation of 5 S complex is temperature-dependent and requires that the estradiol be present as 8 S complex; if the nuclei are exposed to estradiol in cytosol which has been heated to 45°C for 15 minutes to to destroy the 8 S protein, no 5 S complex is observed in the nuclear extract. Independent observations by Brecher *et al.* (1967) that the amount of radioactive estradiol bound to uterine nuclei is greater

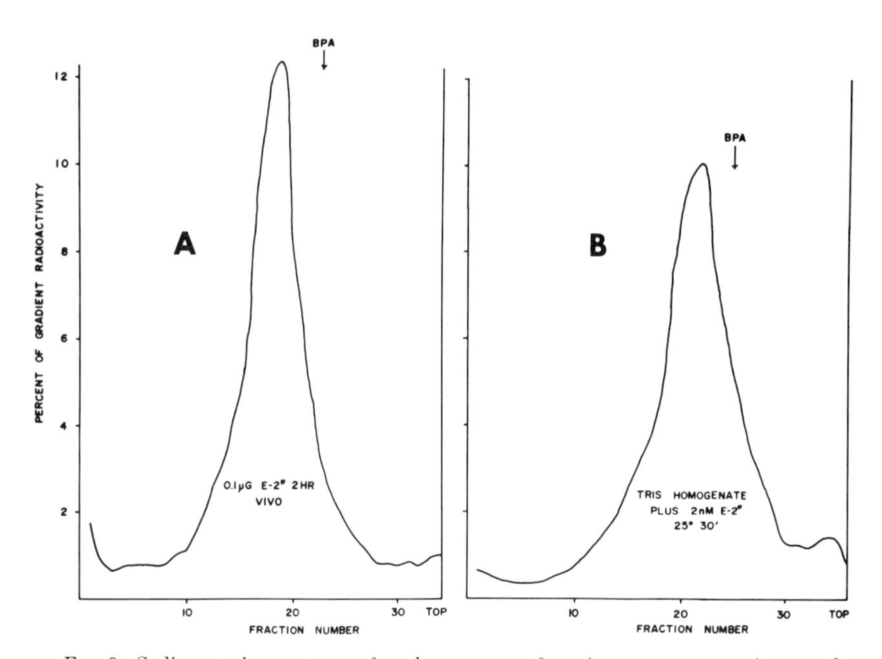

FIG. 9. Sedimentation patterns of nuclear extracts from immature rat uteri exposed to tritiated estradiol: (A) *in vivo* and (B) in the homogenate. Uteri were excised (A) 2 hours after intraperitoneal injection of 0.10 μg of E-2* in 0.5 ml of saline and (B) from untreated rats; the uteri were homogenized in 9 volumes of 10 m*M* Tris, pH 7.4. Homogenate B was diluted with an equal volume of Tris buffer containing E-2* to give a final concentration of 2 n*M*. Both homogenates were incubated for 30 minutes at 25°C and then centrifuged 30 minutes at 25,000 *g*. The washed sediments were extracted with 0.4 *M* KCl in Tris-EDTA, pH 8.5; 200 μl of each extract was layered on 5–20% sucrose gradients, containing 0.4 *M* KCl and Tris-EDTA, pH 7.4 and centrifuged (in separate runs) for 12 hours at 308,000 *g* along with appropriate BPA markers.

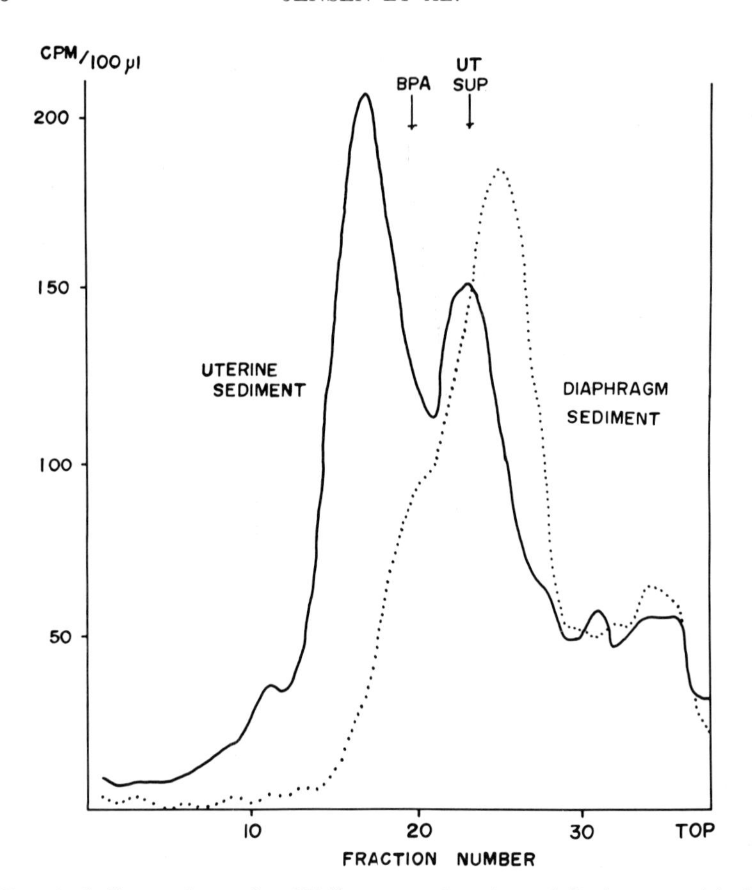

FIG. 10. Sedimentation peaks of KCl extracts of uterine and diaphragm nuclei after incubation with uterine 8 S complex. Uteri and diaphragms from 24-day-old rats were stirred in KRH buffer for 15 minutes at 2°, homogenized in 4 volumes of cold TKE buffer, pH 7.4, and centrifuged for 1 hour at 165,000 *g*. The sediments were rehomogenized gently in the original volume of TKE buffer, and 1.0 ml of each suspension was mixed with 1.0 ml of uterine supernatant which had been diluted with one-fourth volume of E-2* in buffer in a final concentration of 3 n*M*. After incubation for 30 minutes at 25°C, the mixtures were centrifuged for 1 hour at 150,000 *g* and the sediments were extracted with 1.8 ml of 0.4 *M* KCl in Tris-EDTA, pH 8.5. A 200-µl portion of each extract was layered on a 5–20% sucrose gradient containing 0.4 *M* KCl and Tris-EDTA, pH 8.5, and centrifuged 14 hours at 300,000 *g*.

when incubation is carried out in uterine supernatant fraction than in buffer alone, provide additional evidence that the cytosol participates in the incorporation of estradiol into uterine nuclei.

After administration of physiological doses of tritiated estradiol *in vivo*, the extract of uterine nuclei shows only a single 5 S sedimentation peak. Under certain experimental conditions, the extract of the nuclear sediment, after incubation with estradiol in the presence of cytosol, shows only a single peak, similar to that obtained from *in vivo* experiments (Fig. 9). Under other conditions, a significant amount of 4 S complex also is present (Fig. 10). It appears that, in some instances, more 8 S complex can be taken up by the nuclear preparation than can be converted to 5 S complex; this excess 8 S complex is extracted by the KCl as its 4 S subunit. When uterine nuclei are exposed to supernatant fraction in the absence of estradiol, they show no tendency to bind receptor protein; after removal of the supernatant fraction, incubation of such nuclei with estradiol does not give rise to any extractable sedimentation peak.

The foregoing results demonstrate that the 8 S estradiol–receptor complex possesses an affinity for nuclei not shown by the 8 S protein itself. This affinity is not restricted to uterine nuclei. If the nuclear sediments from homogenates of rat diaphragm or liver are incubated with the 8 S complex of uterine cytosol, the uptake of radioactivity by the sediment is nearly as much as that observed with uterine nuclei, but the KCl extracts of these other nuclei do not contain the 5 S complex. In the case of diaphragm nuclei, most of the radioactivity in the extract sediments more slowly than the 4 S subunit of the 8 S complex (Fig. 10); in some experiments a second component is present which sediments slightly faster than the 4 S subunit but slower than the 5 S complex. Although the significance of the products from incubating uterine 8 S complex with nuclei from diaphragm and other nontarget tissues is not yet clear, it appears that only uterine nuclei have the ability to utilize the 8 S estradiol-receptor complex of uterine cytosol to produce the 5 S complex, characteristic of the interaction of estradiol with uterine nuclei *in vivo*.

DISCUSSION AND SUMMARY

The foregoing evidence, obtained by a combination of biochemical and autoradiographic techniques, provides some insight into the nature of hormone-dependent tissues and their interaction with estradiol. During the course of differentiation, the estrogen-dependent tissues not only have lost the ability to grow and function optimally in the absence of the steroid, but they have acquired the capacity to synthesize much larger amounts of a specific extranuclear, sulfhydryl-containing protein than do nontarget tissues. This protein, which shows a striking affinity for estrogenic hormones, has a sedimentation coefficient of about 8 S and is dissociated at high ionic strength into 4 S subunits.

Upon entering the uterine tissue, estradiol appears to associate spontaneously with the 8 S estrophilic protein or "uptake receptor," causing it to interact with the nucleus by a temperature-dependent process which consumes 8 S protein and produces a new entity (Fig. 11). After extraction by potassium-chloride solution, this nuclear complex sediments at about 5 S.

Whether the 8 S complex transfers estradiol to some preexisting nuclear binding site and is destroyed in the process or whether it is itself transformed into the 5 S complex cannot be decided on the basis of present information, although considerable indirect evidence suggests that the 8 S protein actually becomes the 5 S receptor. If this is the case, the 5 S complex might represent the 4 S subunit which has either (1) undergone a conformational change, or (2) lost a buoyant lipid moiety, or (3) become associated with some cellular component by a linkage which, unlike the binding of the

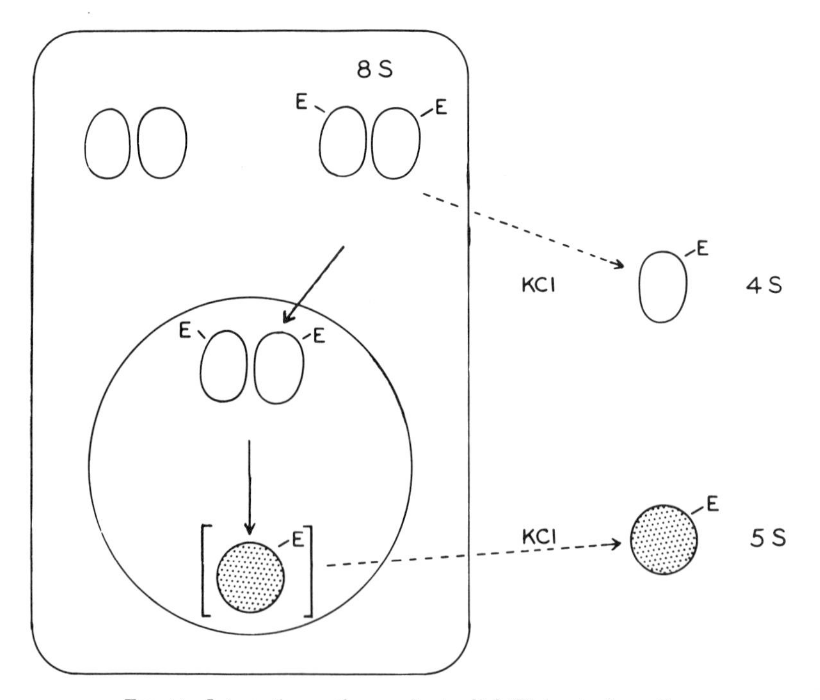

FIG. 11. Interaction pathway of estradiol (E) in uterine cells.

8 S complex to ribonuclease, is not disrupted by salt. Elucidation of the relation between the 4 S and 5 S complexes and their role in growth initiation will be facilitated when tangible amounts of the pure receptor proteins are available for comparison of their composition and structure; isolation of these substances is presently underway (DeSombre *et al.*, 1969).

Meanwhile, as a working hypothesis, we have proposed that the 5 S complex, extracted from the nucleus, consists of the 4 S subunit of the uptake receptor attached to some component of the uterine nucleus which is responsible for the inability of the hormone-dependent tissue to grow and function (by inhibiting gene expression or by some other means). The capture of this inhibitory substance by the hormone-activated receptor protein would relieve the inhibition and allow biosynthetic reactions to commence. Such an indirect process of gene activation, not by the hormone combining directly with the repressor substance but by its activating a specific cellular protein to do so, has been termed "induced derepression" (Jensen *et al.*, 1969a). It provides a definitive and reasonable mechanism for estrogen action which can serve as a guide in designing experiments to test its credibility.

Whatever the exact nature of the intranuclear process which initiates the uterotropic response, one role of the estrogenic hormone appears clear. This function is to associate with the 8 S receptor protein of the cytosol and endow it both with an affinity for the nucleus and the ability to participate in nuclear events leading to growth.

The recent observation of an analogous two-step mechanism for the interaction of dihydrotestosterone with rat prostate (Fang and Liao, 1969; Fang *et al.*, 1969) suggests that both estrogens and androgens may induce growth in their respective target tissues by the same general type of biochemical mechanism. One might speculate on the uniformity of nature's operational pattern by proposing that

When hormones of gonadal extraction
Dance through tissues of sexual attraction,—
  It is not the fandango,
  The twist or the tango,
But the two-step which starts all the action.

ACKNOWLEDGMENTS

These investigations were supported by U. S. Public Health Service research grant CA-02897 from the National Cancer Institute, grant P-422 from the American Cancer Society, and grant 690-0109 from the Ford Foundation.

REFERENCES

Baulieu, E. E., Alberga, A., and Jung, I. (1967). Récepteurs hormonaux. Liaison spécifique de l'oestradiol à des protéines utérines. *Compt. Rend. Acad. Sci. Paris, Ser. D* **265,** 454–457.

Brecher, P. I., Vigersky, R., Wotiz, H. S., and Wotiz, H. H. (1967). An *in vitro* system for the binding of estradiol to rat uterine nuclei. *Steroids* **10,** 635–650.

DeSombre, E. R., Puca, G. A., and Jensen, E. V. (1969). Purification of an estrophilic protein from calf uterus. *Proc. Natl. Acad. Sci. U.S.* **64,** 148–154.

Erdos, T. (1968). Properties of a uterine oestradiol receptor. *Biochem. Biophys. Res. Commun.* **32,** 338–343.

Fang, S., and Liao, S. (1969). Dihydrotestosterone binding by androphilic proteins of rat ventral prostate. *Federation Proc.* **28,** 846.

Fang, S., Anderson, K. M., and Liao, S. (1969). Receptor proteins for androgens: On the role of specific proteins in selective retention of 17β-hydroxy-5-androstan-3-one by rat ventral prostate *in vivo* and *in vitro*. *J. Biol. Chem.* **244,** 6584–6595.

Gorski, J., Toft, D., Shyamala, G., Smith, D., and Notides, A. (1968). Hormone receptors: Studies on the interaction of estrogen with the uterus. *Recent Progr. Hormone Res.* **24,** 45–80.

Hechter, O., and Halkerston, I. D. K. (1964). On the action of mammalian hormones. *In* "The Hormones" (G. Pincus, K. V. Thimann, and E. B. Astwood, eds.), Vol. 5, pp. 697–825. Academic Press, New York.

Jensen, E. V. (1965). Metabolic fate of sex hormones in target tissues with regard to tissue specificity. *Proc. 2nd Intern. Congr. Endocrinol., London, 1964. Excerpta Med. Found. Intern. Congr. Ser.* No. 83, pp. 420–433.

Jensen, E. V., Jacobson, H. I., Flesher, J. W., Saha, N. N., Gupta, G. N., Smith, S., Colucci, V., Shiplacoff, D., Neumann, H. G., DeSombre, E. R., and Jungblut, P. W. (1966). Estrogen receptors in target tissues. *In* "Steroid Dynamics" (G. Pincus, T. Nakao, and J. W. Tait, eds.), pp. 133–157. Academic Press, New York.

Jensen, E. V., DeSombre, E. R., Hurst, D. J., Kawashima, T., and Jungblut, P. W. (1967). Estrogen-receptor interactions in target tissues. *Colloq. Intern. Physiol. Reprod. Mammiféres, Paris 1966* (A. Jost, ed.), *Arch. Anat. Microscop. Morphol. Exptl.* **56** (Suppl.), 547–569.

Jensen, E. V., Suzuki, T., Kawashima, T., Stumpf, W. E., Jungblut, P. W., and DeSombre, E. R. (1968). A two-step mechanism for the interaction of estradiol with rat uterus. *Proc. Natl. Acad. Sci. U.S.* **59,** 632–638.

Jensen, E. V., Suzuki, T., Numata, M., Smith, S., and DeSombre, E. R. (1969a). Estrogen-binding substances of target tissues. *Steroids* **13,** 417–427.

Jensen, E. V., DeSombre, E. R., Jungblut, P. W., Stumpf, W. E., and Roth, L. J. (1969b). Biochemical and autoradiographic studies of ^{3}H-estradiol localization. *In* "Autoradiography of Diffusible Substances" (L. J. Roth and W. E. Stumpf, eds.), pp. 81–97. Academic Press, New York.

Jungblut, P. W., Hätzel, I., DeSombre, E. R., and Jensen, E. V. (1967). Über Hormon-"Receptoren;" die oestrogenbindenen Prinzipien der Erfolgsorgane. *In* "Wirkungsmechanismen der Hormone." *Colloq. Ges. Physiol. Chem.* **18,** 58–86.

King, R. J. B., and Gordon, J. (1966). The localization of [6,7-^{3}H]-oestradiol-17β in rat uterus. *J. Endocrinol.* **34,** 431–437.

King, R. J. B., Gordon, J., and Inman, D. R. (1965). The intracellular localization of oestrogen in rat tissues. *J. Endocrinol.* **32,** 9–15.

KORENMAN, S. G., and RAO, B. R. (1968). Reversible disaggregation of the cytosol-estrogen binding protein of uterine cytosol. *Proc. Natl. Acad. Sci. U.S.* **61,** 1028–1033.

NOTEBOOM, W. D., and GORSKI, J. (1965). Stereospecific binding of estrogens in rat uterus. *Arch. Biochem. Biophys.* **111,** 559–568.

PUCA, G. A., and BRESCIANI, F. (1968). Receptor molecule for estrogens from rat uterus. *Nature* **218,** 967–969.

ROCHEFORT, H., and BAULIEU, E. E. (1968). Récepteurs hormonaux: relations entre les "récepteurs" utérins de l'estradiol. *Compt. Rend. Acad. Sci. Paris, Ser. D* **267,** 662–665.

ROCHEFORT, H., and BAULIEU, E. E. (1969). New *in vitro* studies of estradiol binding in castrated rat uterus. *Endocrinology* **84,** 108–116.

SHYAMALA, G., and GORSKI, J. (1969). Estrogen receptors in the rat uterus. *J. Biol. Chem.* **244,** 1097–1103.

STUMPF, W. E., and ROTH, L. J. (1966). High resolution autoradiography with dry mounted, freeze-dried frozen sections. *J. Histochem. Cytochem.* **14,** 274–287.

STUMPF, W. E. (1968). Subcellular distribution of ^{3}H-estradiol in rat uterus by quantitative autoradiography: A comparison between ^{3}H-estradiol and ^{3}H-norethynodrel. *Endocrinology* **83,** 777–782.

TALWAR, G. P., SEGAL, S. J., EVANS, A., and DAVIDSON, O. W. (1964). The binding of estradiol in the uterus: A mechanism for derepression of RNA synthesis. *Proc. Natl. Acad. Sci. U.S.* **52,** 1059–1066.

TOFT, D., and GORSKI, J. (1966). A receptor molecule for estrogens: Isolation from the rat uterus and preliminary characterization. *Proc. Natl. Acad. Sci. U.S.* **55,** 1574–1581.

TOFT, D., SHYAMALA, G., and GORSKI, J. (1967). A receptor molecule for estrogens: Studies using a cell-free system. *Proc. Natl. Acad. Sci. U.S.* **57,** 1740–1743.

DEVELOPMENTAL BIOLOGY SUPPLEMENT 3, 172–205 (1969)

The Action of Auxin on Cell Enlargement in Plants

PETER M. RAY

Department of Biological Sciences, Stanford University, Stanford, California

INTRODUCTION

The plant growth substance auxin is the classical botanical example of a communication carrier between cells. The occurrence and consequences of this communication, as with any hormonal communication system, depend (1) upon regulation of the formation and the inactivation of the growth substance; (2) upon the mechanism of its transport, which in the case of auxin is highly directional and specific for this class of substances; and (3) upon the nature of its action on target cells, which varies widely in different cells and tissues and thus permits a wide variety of "messages" to be transmitted by a single hormone. We are interested in the present article only in this latter aspect of auxin physiology, and more specifically with the mechanism of the auxin effect on cell enlargement, which is the classical (but by no means the only) action of auxin upon target cells.

In the immediate mechanism by which a plant cell grows in size, we are dealing essentially with the rate at which its cell wall stretches under turgor stress and cell volume consequently increases by osmotic absorption of water. It follows that any agent that influences growth must act by influencing, directly or indirectly, the mechanism of cell wall expansion or its driving force, the turgor stress. Although it was thought for a time in the 1930's that auxin might act directly upon the cell wall to affect its extensile properties, this idea has long since been abandoned in view of the dependence of cell enlargement upon cellular metabolism as shown by inhibitor experiments. Cell wall enlargement is evidently not a simple physical stretching and it seemed likely that auxin must act on it by influencing biochemical processes within the cell, upon which the process of wall expansion depends.

This means that the growth response to auxin involves an additional *intracellular* communication phenomenon; namely, between the site of direct action of auxin within the cell, and the cell wall to the outside. This view is reinforced by high-resolution measurements of the timing of the elongation response to auxin which show the existence of a latent period (Fig. 1), lasting about 10 minutes at 25°C, between treatment with auxin and onset of accelerated elongation rate in coleoptiles (Ray and Ruesink, 1962) and stem segments (Barkley and Evans, 1969; Uhrström, 1969). Treatment under reduced temperature or in presence of KCN at low concentration extends the latent period (Evans and Ray, 1969), presumably by inhibiting the metabolic chain of events that leads ultimately to an action on the cell wall. Rather similar timing has been measured for the inhibitory effect of auxin on root elongation (Hejnowicz and Erickson, 1968; Bottrill and Hanson, 1968).

To achieve an explicit understanding of how auxin influences growth, we must seek to discover the nature of its effect on the cell wall, the chain of events that precedes this effect, and the intracellular process or property that sets the chain in motion and is directly influenced by auxin.

By comparison with the action of other growth hormones in plants or animals the auxin response, in which the full effect on growth rate is attained within a few minutes, is remarkably rapid. The present article will consider how the timing of the auxin response bears on current hypotheses regarding the biochemistry and biophysics of the cell wall expansion process and the action of auxin thereon.

A distinction should first be made between whether the hormone acts directly on the cellular machinery that supports the growth

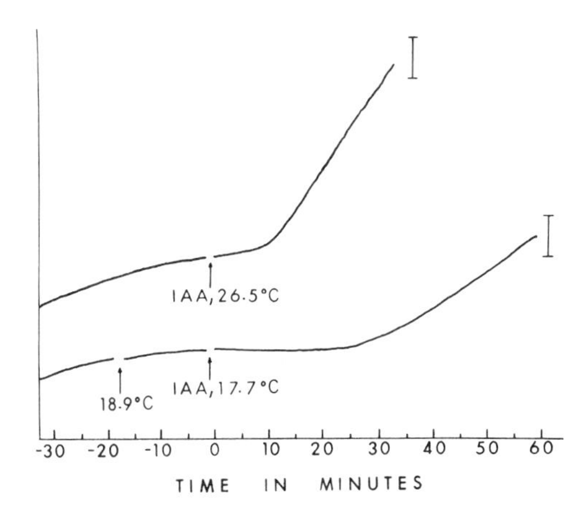

FIG. 1. Timing of the response of *Avena* coleoptile segments to treatment with 3 μg/ml indoleacetic acid (IAA) at 26.5°C and at 17.7°C. The vertical bar represents 1 mm of total elongation by thirteen segments initially 8 mm long. From Evans and Ray (1969).

process or whether it influences the production of an effector molecule that in turn controls the machinery of the growth process. The latter type of mechanism is illustrated by hormonal responses that are mediated by cyclic 3′,5′-adenosine monophosphate in animal tissues. In the case of auxin, one class of responses clearly falls into this category, namely, inhibitions of cell elongation that are actually responses to ethylene, the biogenesis of which is stimulated by auxin (Burg and Burg, 1968). This effect has been invoked to explain the inhibition by auxin of growth of roots, buds, hypocotyl hook cells, and of stem tissue at high doses of auxin. The means by which ethylene inhibits cell elongation is not definitely known in any of these responses, but evidently ethylene must have an effect (not necessarily direct) on the mechanisms that control the rate and directionality of cell wall expansion.

Promotion of cell enlargement by auxin is not, in general, attributable to ethylene nor, so far as is definitely established, to other intermediary effectors. However, a hypothesis has been advanced that auxin effects on growth are due actually to oxidation products derived from the hormone (Moyed and Tuli, 1968). This hypothesis seems dubious for a number of reasons, including the high auxin activity of synthetic analogs that are not oxidized by tissue, and is not supported by quantitative evidence (Andreae and Collet, 1968).

GENE ACTIVATION HYPOTHESIS

Much effort has lately been directed at demonstrating promotions by auxin of RNA and protein synthesis, and inhibitions of auxin-induced growth by inhibitors of RNA and protein synthesis (Key, 1969). These effects have been widely interpreted as evidence that auxin acts on growth by promoting specific protein synthesis via a promotion of transcription, a now popular hypothesis for hormone action generally.

The weakness of the inhibitor experiments is, of course, that these agents will affect growth if growth itself or auxin action depends in any way upon RNA and protein synthesis, whether or not auxin acts on growth by stimulating transcription.

The promotive effects of auxin on incorporation of labeled precursors into RNA and protein have been detected mostly after periods of about an hour or more, which seems much too slow to account for the characteristic rapid promotion of elongation by auxin illustrated in Fig. 1. Double-label experiments designed to detect differences in incorporation into specific proteins due to auxin (Patterson and Trewavas, 1967) indicated a multitude of differences rather than the specificity that is hoped for under the gene activation hypothesis.

These late auxin effects on RNA and protein synthesis may well be of importance in longer-term auxin responses, but to relate them to the rapid cell elongation response it seems necessary to construe them as late reflections or amplifications of a much more specific rapid effect on transcription. Such an explanation is biologically somewhat implausible since, as illustrated in Fig. 1, the elongation rate response to auxin is maximal after little more than 10 minutes, whereas the wide variety of enzyme induction and protein synthesis responses to hormones that are known in animal systems, as well as those demonstrated in plants, occur over a time scale of hours to days (cf. references cited by Evans and Ray, 1969). However, this might mean merely a much more direct action of auxin than of other hormones upon transcription. A rapid (10–15 minutes) effect of auxin on incorporation into RNA has been claimed recently for two plant systems (Matthysse, 1968; Masuda and Kamisaka, 1969).

We tested the gene activation hypothesis of auxin action by examining the effects of inhibitors of RNA and protein synthesis upon the timing of the response (Evans and Ray, 1969). This test rests on the principle that the time constants for response of the epigenetic system must be long compared with those of other metabolic systems with which it is coupled, and hence must dominate the timing of the physiological response if it is due to gene activation. Initially we expected that if elongation results from auxin-induced transcription and translation, the latent period of the elongation response should be lengthened by inhibitors of these processes. We found, however, that doses of actinomycin D, cycloheximide, or puromycin that caused substantial inhibition of auxin-induced elongation did not extend the latent period, and cycloheximide actually shortened it (see Fig. 2). Very similar observations have been made on pea stem segments (Barkley and Evans, 1969).

Kinetic analysis indicated that the results were compatible with the gene activation hypothesis only on the assumption that the half-life of the auxin-induced mRNA and protein is a small fraction of 10 minutes. This assumption is very improbable on the basis of the extensive evidence regarding RNA and protein turnover in tissues of higher organisms, except possibly for turnover of RNA in the nucleus, which, however, is not kinetically relevant since the timing

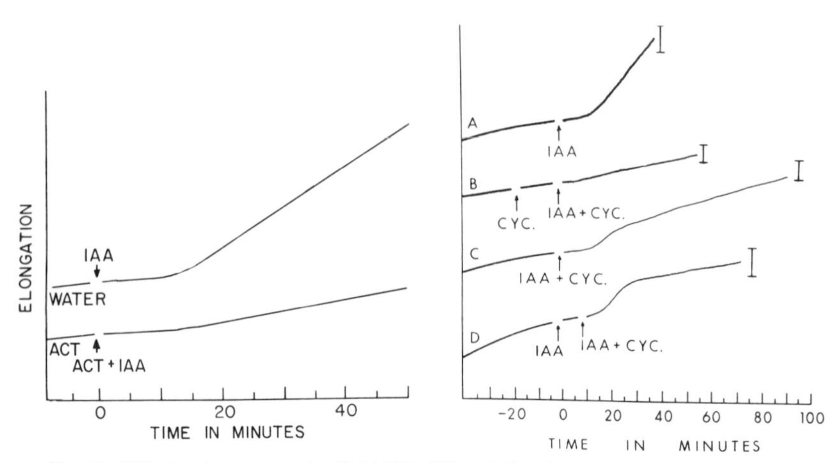

FIG. 2. Effects of actinomycin D (*ACT*) (20 μg/ml, 1 hour pretreatment) and cycloheximide (*CYC.*) (3 μg/ml) upon timing of auxin response of *Avena* coleoptile segments. IAA = indoleacetic acid. From Barkley and Evans (1969) and Evans and Ray (1969).

must depend upon cytoplasmic mRNA turnover if cytoplasmic ribosomes are responsible for the translation step. The assumption was, moreover, contraindicated by the data themselves since actinomycin D inhibited elongation only very slowly when added after auxin.

Evans and Hokanson (1969) and Hertel *et al.* (1969) have extended this point by determining the time course of inhibition when auxin-stimulated coleoptile tissue is treated with the competitive auxin antagonist 4-chlorophenoxyisobutyric acid. Inhibition became fully established within 20 minutes; this indicates that none of the products of auxin action that lie on the pathway to cell wall expansion can have a half-life of more than a few minutes (Fig. 3).

These observations cast serious doubt on the gene activation hypothesis, at least as regards the auxin effect on elongation in coleoptiles and stem segments. In order to maintain the hypothesis it would be necessary, not only to demonstrate a substantial auxin effect on synthesis of some RNA and protein species in this material within 10 minutes, but also to demonstrate that these species possess the required turnover characteristics. To prove the hypothesis it would further have to be shown that the protein in question can and does lead within 10 minutes to an effect on the cell wall capable of causing a maximal rate of expansion.

Barkley and Evans (1969) noted the curious fact that during the latent period of auxin action the rate of elongation is actually *depressed* below that which prevailed before exposure to auxins. This virtually immediate, inhibitory effect of auxin on elongation may well be exerted at the outside of the cell membrane. In the case of pea stem and cucumber hypocotyl tissue this period of inhibition is followed by a transient rapid elongation that much exceeds the steady, auxin-promoted elongation rate which is attained after about half an hour. The net effect of these events is to create, in long-term measurements of elongation (intervals of 0.5 hour or more), the appearance that auxin promotion of elongation occurred virtually at the moment of exposure to auxin.

One possible interpretation of these observations is that the auxin action that is promotive for elongation does actually occur virtually at the time of exposure to auxin, but its manifestation in the growth rate is temporarily suppressed by a separate, inhibitory effect that operates when the auxin concentration outside the cell membrane is much greater than that inside.

Evans has recently discovered that if maize coleoptiles are cut into short (4 mm long) segments to maximize the rate of auxin uptake and are then treated with a low (10^{-7} *M*) concentration of indoleacetic acid to minimize the initial inhibitory effect of external auxin, their elongation is promoted maximally by auxin in 2 minutes (Fig. 4). This finding indicates that the elongation response to auxin can be

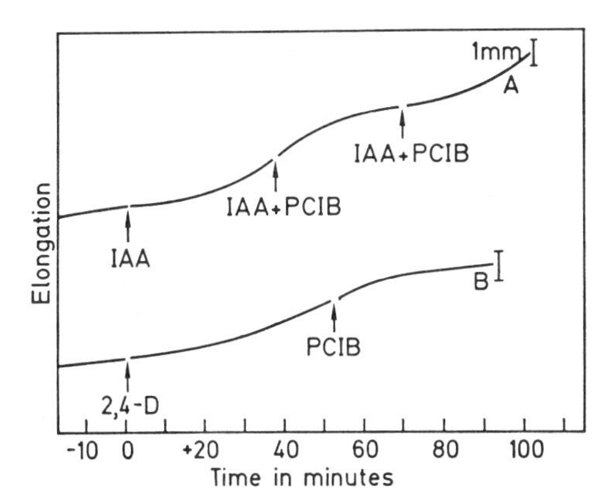

FIG. 3. Effect of 10^{-4} *M* 4-chlorophenoxyisobutyric acid (PCIB) upon elongation of *Zea mays* coleoptile segments in 10^{-6} *M* indoleacetic acid (IAA) or 2,4-dichlorophenoxyacetic acid (2,4-D). At the third arrow in curve *A* the IAA concentration was raised to 5×10^{-4} *M*. From Evans and Hokanson (1969).

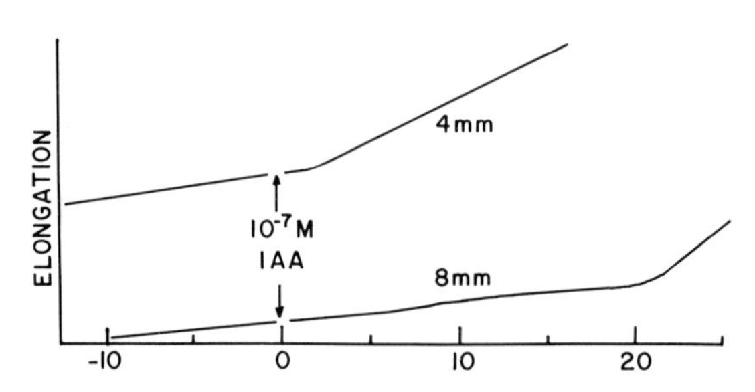

FIG. 4. Elongation response to 8 mm-long and 4 mm-long segments of *Zea mays* coleoptiles to 10^{-7} *M* indoleacetic acid (IAA). Unpublished data of M. L. Evans.

virtually immediate, and it increases further the stringency of the conclusions regarding the gene activation hypothesis just discussed.

ALTERNATIVE MODES OF AUXIN ACTION

Cycloheximide was found to cause complete inhibition of coleoptile elongation in about 20 minutes (Evans and Ray, 1969). This agreed with rapid effects of cycloheximide on growth as reported in various other systems, without employment of methods for following the time course precisely. The kinetic observations might be compatible with auxin regulation of translation as the basis of the elongation response, provided that either (a) the auxin-induced protein has a half-life of not more than a few minutes (improbable as mentioned above), or (b) the protein is employed in growth soon after its formation but not subsequently—for example, as a structural protein rather than a catalyst.

MacDonald and Ellis (1969) have recently reported that cycloheximide can inhibit energy metabolism in plant tissues. This casts some doubt upon the conclusion, from experiments with cycloheximide, that cell enlargement depends very immediately upon protein synthesis, since inhibitors of energy metabolism such as cyanide inhibit elongation very rapidly (Ray and Ruesink, 1962).

The time required for the complete elongation rate response to auxin is dangerously near (or, in the case of the results in Fig. 4, less than) the time estimated to be required to assemble a single polypeptide chain in higher organisms (Haschemeyer, 1969, and references there cited). It is difficult to believe that a steady state in respect not only to the relevant protein but to auxin uptake and its promotive effect on protein synthesis on the one hand, and interaction of the supposed protein with the cell wall on the other, could be achieved in a time similar to that required for the first protein molecules to be completed. Therefore, I consider that auxin-induced synthesis of proteins constitutes at present a very dubious hypothesis as an explanation of the rapid auxin action on cell elongation. Its substantiation would require attention to the same points specified in concluding the discussion of the gene activation hypothesis.

Preoccupation with the idea that auxin must regulate growth via an effect on specific protein synthesis (via transcriptional or translational control) has virtually eclipsed serious investigation of alternative explanations in the last few years, and I feel that the time is ripe for such investigation. The two principal classes of action to be distinguished are effector action on enzymes, and interaction with membranes leading to effects on transport between cell compartments. Some effects of the former sort have been claimed recently (Mitchell and Sarkissian, 1966) but seem dubious (Zenk and Nissl, 1968; Sarkissian and Schmalstieg, 1969) and unrelated to the process of cell wall expansion.

Ideas of auxin interaction with membranes were formerly entertained (Veldstra, 1953), and are logical in terms of the combination of hydrophilic and bulky hydrophobic groups that is characteristic of molecules active as auxins, and the relatively nonspecific structural requirements for auxin activity. It seems inevitable that molecules of the auxin type will enter and orient themselves in membranes, as has been demonstrated with spin-labeled fatty acids and steroid derivatives (Hubbell and McConnell, 1968, 1969). Such interactions of auxins with membranes have been visualized (van Overbeek, 1959, 1961) but not directly detected, although the adsorption of auxins by lipids has been demonstrated (Brian and Rideal, 1952; Weigl, 1969). The virtually immediate effects of auxin on elongation rate found in some situations, as discussed above, seem understandable most readily in terms of an action at the cell surface.

Some limited effects of auxins on membrane potentials have been recorded (see discussion by Jaffe in this volume). One type of transport effect of auxin is definitely established, namely, the promotion of its own uptake (Poole and Thimann, 1964) and polar transport (Hertel and Flory, 1968). This effect could result from the promotion of protoplasmic streaming by auxin (Sweeney and Thimann, 1938) rather than from an action at cell membranes. Promotion of protoplasmic streaming is the one action of auxin that has been observed to occur as rapidly as, or more rapidly than, the effect of auxin on elongation. The means by which auxin causes the effect is not known, nor has a direct connection between this effect and the cell enlargement process been demonstrated. However, most of the models for auxin action on elongation to be discussed below could rest upon an accelerated redistribution of substrates or enzymes within the cell, an effect that could well be brought about by an increase in streaming rate.

Perhaps the simplest conceivable transport effect of auxin in relation to cell wall growth would be to promote efflux of an enzyme, a structural polymer, or a metabolite through the plasma membrane into the cell wall, where it is used in the process of wall expansion. The biosynthesis of the metabolite or polymer within the cell would be accelerated in turn by well-known mechanisms of feedback control, and this would be the part of the mechanism that depends upon RNA and protein synthesis.

Some indications of an effect of auxin on transport of material into the cell wall will be discussed below.

PHYSICAL NATURE OF THE GROWTH RESPONSE TO AUXIN

Early in the history of auxin research it was recognized that auxin treatment modifies the mechanical behavior of the cell wall in such a way as to lead to a more rapid rate of wall expansion. This was termed an increase in "plasticity" of the cell wall. Under the gene activation hypothesis it is generally anticipated that auxin action induces formation of enzymes that are capable of increasing the plasticity. As mentioned above, the release of wall-softening enzymes could of course be involved even if auxin is not acting by inducing transcription or translation.

Much effort has been expended in the last few years to replace the previously nebulous and physically ill-founded concept of wall plasticity with a proper biophysical understanding of the mechanics of cell wall expansion during growth. Lockhart (1965) has reviewed the elementary principles of the subject; his article may be consulted for definitions of concepts and terms that are employed here.

Two principal approaches to the problem have been made thus far: (a) mechanical extension measurements on cell-wall specimens, and (b) high-resolution elongation measurements of the type discussed above, whereby the immediate response of strain rate to changes in stress and other relevant parameters may be measured using living cells.

Effects on Elongation Rate

Osmotic experiments on growing cells and tissues show that elongation rate drops rapidly as turgor stress is diminished, and falls to zero at a still substantial value of turgor (Cleland, 1959; Ray and Ruesink, 1963; Green, 1968). This indicates the existence of a plastic (in the physical sense) component in the wall expansion process; that is, a mechanism in which irreversible strain occurs only when the stress exceeds a yield value, presumably determined by definite bonding forces within the wall material. The yield stress was found to be substantially lower in auxin-treated coleoptile tissue than in controls (Fig. 5). This suggests that auxin treatment may cause a reduction in the density (frequency) of the bonding interactions that are responsible for the yield stress effect.

A somewhat different picture of the auxin effect was obtained in elongation measurements that were made after a period of hours (Cleland, 1959, 1967b). The results indicated existence of a "critical stress" below which auxin does not promote cell enlargement. These experiments involved the combined effect of turgor differences on elongation rate and on the action of auxin on the cell wall, as well as the possibility of secondary changes in rate following alterations in turgor; so they should not be compared directly with the foregoing.

Treatment of coleoptile segments with Ca^{2+} raised the yield stress (Fig. 6) which implies that Ca^{2+} may participate in ionic bonding interactions that are involved in the passive yielding properties of the cell wall.

Results of temperature, inhibitor, and oxygen pulse experiments (Ray and Ruesink, 1962) indicated that the immediate strain rate in growing oat coleoptile cells is not determined purely by a passive viscous retardation property but involves a metabolic (chemorheological) component. Elongation rate is reduced immediately, with a very high Q_{10} (ca. 3.5) when temperature is lowered, and it falls rapidly (and reversibly) in low concentrations of KCN and rises rapidly (within 2–3 minutes) when oxygen is supplied to O_2-depleted tissue.

A very curious phenomenon is seen when oat coleoptile tissue is treated with high concentrations of CO_2 or with acidic media of about pH^4 (Fig. 7). Within 1 minute the elongation rate jumps to a value several times greater than is attainable by treating with auxin. In the case of CO_2 treatment this accelerated elongation continues for only about half an hour and a much weaker response is obtained upon a second exposure to CO_2, whereas in buffered media the ac-

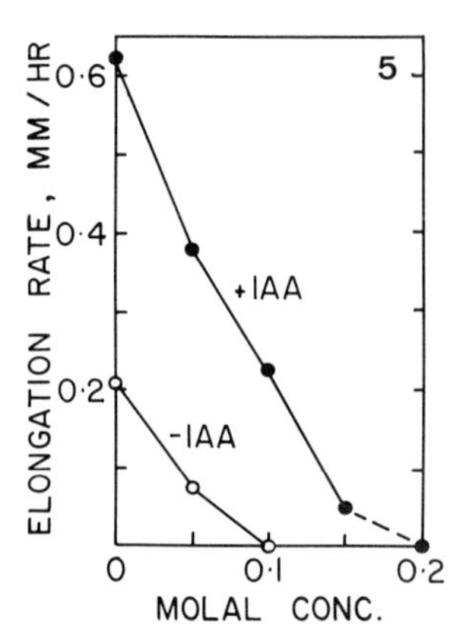

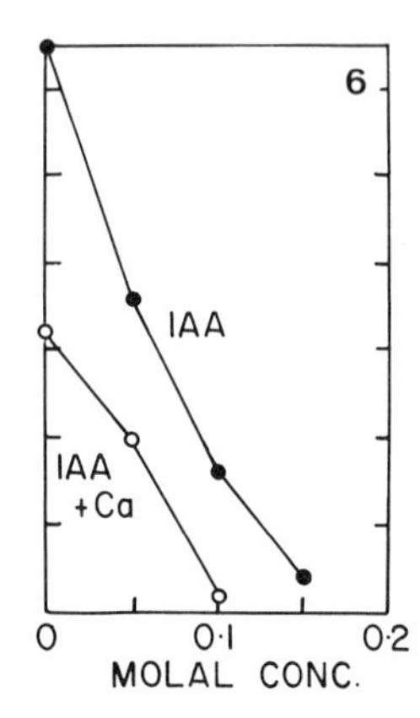

FIG. 5. Immediate effect of mannitol upon elongation rate of *Avena* coleoptile segments pretreated with water or with 3 μg/ml indoleacetic acid (IAA) for 20 minutes. From Evans (1967).

FIG. 6. Immediate effect of mannitol on elongation rate of *Avena* coleoptile segments in IAA and in IAA plus 10^{-3} *M* $CaCl_2$. From Evans (1967).

celerated strain rate persists for at least 90 minutes. This rate cannot be accelerated further by auxin. The most remarkable thing about it is that metabolic inhibitors such as cyanide and mercurials, which rapidly suppress normal elongation either in presence or absence of auxin, have no effect whatever on the elongation induced by CO_2 or acidic media. This elongation is inhibited, however, by mannitol at concentrations that osmotically inhibit normal growth. The CO_2 or acid-induced elongation thus appears to be a purely passive, physical strain process. Transfer of CO_2-treated tissue to a neutral buffered medium does not reverse the CO_2-induced elongation rate.

This phenomenon may be extremely significant in relation to the rheology of normal growth and the auxin effect thereon. Reinhold and Glinka (1966) described an apparently similar effect of CO_2 on several tissues. The simplest explanation seems to be that acidification disrupts bonding forces between mechanical elements of the wall structure that normally restrict strain and that are acted on chemorheologically, during normal growth, by the process that is promoted by auxin. One would suppose, in view of the pH value that is effective, that ionic interactions between acidic polysaccharides (polyuronides) are being eliminated by acidification. There is evidence that Ca^{2+} retards the phenomenon. The effect of CO_2 and acid suggests that normal growth does not involve the extension of a covalently cross-linked network structure.

Mechanical Extension Experiments

In extensiometer experiments (Cleland, 1967a; Lockhart, 1967) performed on wall material from growing plant tissue, stress is observed to rise continuously with strain, and the stress at any strain during the first extension behaves as a yield stress. Therefore, yield stress appears to rise approximately linearly with strain; this relation has been regarded as "strain hardening." The slope of this irreversible strain/yield stress relation is referred to as plastic extensibility. The principal effect of auxin that is detected in extension experiments is an increase in this plastic extensibility.

The relation between this effect and the growth process is a problem for several reasons: (1) Turgor stress remains constant during steady cell enlargement and, if anything, decreases (for osmotic reasons) when auxin promotes cell enlargement, whereas the plastic extensibility pertains to an increase of strain with *increasing* stress. (2) Since this strain/stress relation is virtually independent of time, plastic extensibility and turgor stress cannot by themselves determine a strain *rate* nor, therefore, the rate of cell enlargement. (3) The immediate dependence of strain rate on turgor stress found osmotically with growing coleoptile tissue (Ray and Ruesink, 1963; Evans, 1967) shows that its cell walls do not, like the wall specimens in extension measurements, lie virtually at a yield stress during steady straining.

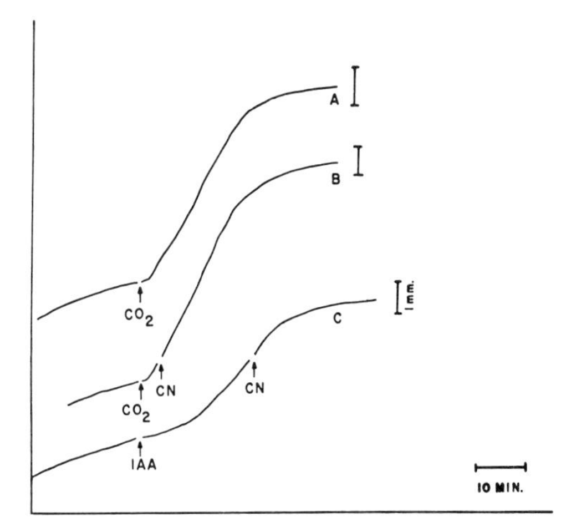

FIG. 7. Effect of treatment with CO_2-saturated water, on elongation of *Avena* coleoptile segments, as compared with effect of IAA. At the arrow labeled *CN*, medium was replaced by 10^{-3} *M* KCN. From Evans (1967).

(4) The auxin-induced plastic extensibility appears to be "used up" or converted into wall extension during the growth that is supposed to result immediately from this extensibility (Cleland, 1968a), whereas the difference between the measured strain-hardening curves of wall material from control and auxin-treated tissue does not disappear in the course of a large amount of straining in the extensiometer test.

Therefore, parameters additional to, or other than, plastic extensibility must be involved in determining the strain rate as a function of turgor stress during growth. This may explain why plastic extensibility correlates poorly with growth rate, in various instances; for example, under treatment with agents such as inhibitors of RNA and protein synthesis or as a function of time after auxin is supplied or withdrawn (Cleland, 1965; Morrè and Eisinger, 1968; Cleland, 1968c). In order to obtain a workable growth mechanism we are obliged to assume that growth depends upon the occurrence of some kind of metabolic stress relaxation phenomenon. This was deduced earlier on kinetic grounds as discussed above (Ray and Ruesink, 1962). The plastic extensibility that is measured in artificial extension experiments may be taken to reflect the operation of the passively yielding mechanical elements in the cell wall that determine, *in vivo*, how much strain will occur per unit of metabolically caused stress relaxation. Cleland (1968a) has visualized the growth process as a continuing sequence of metabolic "wall loosening" events, each followed by a limited amount of plastic straining set by the plastic extensibility.

The principal question of importance to the auxin field is then whether auxin acts on the growth rate by influencing the rate of metabolically caused stress relaxation or by influencing the physical properties that comprise the measured plastic extensibility, if indeed it acts on only one of these. There is ample evidence, despite the discrepancies mentioned above, that the growth process is related to the auxin effect on plastic extensibility; that the latter effect is not merely a result of occurrence of cell enlargement (Cleland, 1968a); and that it too depends upon metabolism. If we attribute all the effect of auxin on growth to its influence on plastic extensibility, we can account for discrepancies between the effects of various agencies on growth rate and on extensibility by supposing that some of these agencies affect the rate of metabolic stress relaxation. For example, cycloheximide was found to have no effect on extensibility of coleoptile tissue previously treated with auxin (Cleland, 1968c), even though elongation rate is suppressed rapidly by this inhibitor (Evans and Ray, 1969). We may thus conclude that protein synthesis is required for occurrence of metabolic stress relaxation, and not for the continued existence of high plasticity (although it is perhaps required for the auxin induction of increased plastic extensibility; cf. Black *et al.*, 1967, and other references cited above).

One type of discrepancy cannot, however, be so explained—namely, between the timing of the elongation rate response to auxin, discussed earlier, on the one hand and of the plastic extensibility response to auxin on the other. All measurements of the latter agree in showing that it occurs over a time scale of one to a few hours, whereas if it governed the growth rate in response to auxin it should be completed within about 10 minutes. Also troublesome is the fact that in no case does the factor by which plastic extensibility increases under auxin treatment approach anywhere near the factor by which the growth rate is increased by auxin, and this holds *a fortiori* for the early period of auxin action, before maximum plasticity increase has been reached but when growth rate is already maximum. A similar lack of time congruence is seen in an early auxin effect on elastic deformability of sunflower hypocotyl tissue detected by Uhrström (1969) using resonance measurements. Masuda (1969) claimed to detect a rapidly completed (15 minutes) auxin effect on elastic extensibility of coleoptiles, but the reported effect was extremely small.

Therefore, we are forced to choose between the alternatives, either that the measured plastic (or elastic) extensibility bears only a remote and variable relationship to the physical properties that actually determine wall expansion *in vivo* as a function of metabolic stress relaxation, or else that auxin must accelerate the latter process. Since the measured plastic extensibility is itself demonstrably dependent upon both auxin and metabolism, the most economical hypothesis would be that auxin promotes the metabolic stress relaxation process and that operation of this process as well as of others such as the actual straining, and also the process of wall deposition that normally accompanies it, all affect the measured extensile properties of the cell wall and contribute to the aforementioned discrepancies between growth rate and extensibility. This view of auxin action accords with much evidence; for example, it explains the observation that the auxin effect on plastic extensibility is suppressed by reducing the turgor (Cleland, 1967b), since it is essential both phenomenologically and in principle to assume that the rate of a chemical stress relaxation mechanism is proportional to the stress borne by it.

The alternative view, that auxin does not accelerate the metabolic stress relaxation process but only influences physical parameters that determine the stress/irreversible strain relation, can be maintained by supposing that plastic extensibility is no longer "rate-limiting" under auxin-promoted conditions. One encounters difficulty in formulating in physical terms what is meant by "not rate-limiting"; moreover, this view also compels us to make the assumption that most of the plastic extensibility that is seen in extensiometer experiments is concerned with strain mechanisms that do not operate during growth. Kinetic experiments upon oat coleoptiles do, however, give indication that the auxin-promoted growth rate is limited by a parameter different from that which controls growth rate in the absence of auxin (Ray, 1961).

These alternative views of the biophysics of auxin action on cell enlargement have important implications with regard to the timing characteristics to be expected for the biochemical action of auxin on the cell wall that is relevant to growth rate, and therefore with regard to the criteria by which such a mechanism should be recognized and critically tested. Unfortunately, it is not yet possible, in my opinion, to distinguish conclusively between these two classes of rheological influences of auxin, and I feel that this should be an important goal for future work.

BIOCHEMICAL ACTION ON THE CELL WALL

Three principal classes of biochemical processes can be conceived to be involved in the extensile properties of the cell wall and the effect of auxin thereon: (1) splitting or degradation of polymer chains, (2) splitting of cross-links between the chains, and (3) biosynthesis of new wall material.

Cross-links

Of the considerable variety of possible chemical cross-links that can be imagined between the polysaccharides of plant cell walls, as illustrated in a diagram published by Preston and Hepton (1960), the only type for which any appreciable experimental support seems to exist, is divalent cation cross-links between acidic groups (uronic acids). This was the basis for a theory of wall growth and auxin action by Ordin *et al.* (1957): methyl esterification of polyuronide carboxyl groups would split such ionic cross-links, leading to strain; auxin treatment was found to promote incorporation of methyl ester groups into the cell wall. This theory was given up primarily because of findings by Cleland (1963a,b) that the promotion of esterification by auxin is of restricted occurrence and can be prevented without preventing the auxin effect on growth, and that the expected promotion of isotopic Ca^{2+} exchange in auxin could not be observed (Cleland, 1960). Furthermore, the polyuronide methylating system is associated with the cytoplasmic organelle that synthesizes polysaccharides, and methylation is performed upon recently synthesized polymer prior to its export into the cell wall (Kauss and Swanson, 1969), rather than upon previously deposited polyuronides as required by the ionic cross-link splitting hypothesis.

Lamport (1965) has developed evidence that the plant cell wall contains a distinctive class of protein, rich in hydroxyproline, which is a structural protein, and likely to be important in the mechanism of growth—for example, by serving as a labile cross-link between wall polysaccharides. He named the hydroxyproline-rich protein (hypro-protein) of the cell wall "extensin" in anticipation of this function.

In the bacterial wall peptidoglycan, the polysaccharide chains are extensively cross-linked by oligopeptide chains (Strominger *et al.*, 1967), to form a network structure. This structure can undergo strain and permit the cell to grow only by severance of the cross-links or of the glycosaminoglycan backbones (Schwarz *et al.*, 1969).

Because of their lack of amino sugars the polysaccharides of higher plant walls cannot be cross-linked by oligopeptide chains in the manner of bacterial peptidoglycan. Lamport (1967, 1969) has found that oligosaccharide or polysaccharide chains are linked to hypro-protein by joining an arabinose unit glycosidically to the OH group of hydroxyproline. Since each oligosaccharide or polysaccharide chain can be linked just once (at its reducing end) in this manner this permits building of supramolecular aggregates but not of a true cross-linked network structure in the rheological sense, which requires a minimum of two cross-links per component polymer chain.

Lamport (1965) suggested that hypro-protein chains may be cross-linked by disulfide bridges. This could not create a cross-linked polysaccharide network structure, but could yield a hypro-protein network bearing polysaccharide side chains. A disulfide cross-linked model, involving glucomannan-protein complexes as the structural units, had been proposed by Nickerson (1963) for the cell wall of the yeast *Candida*. He visualized the mechanism of wall growth as depending upon a reductive splitting of the disulfide cross-links by a specific protein, disulfide reductase. It was not directly demonstrated for yeast that the wall expands in this manner, and no evidence for such a mechanism has yet been forthcoming for higher plants, although Thompson and Preston (1968) reported some indication that disulfide bridges might have a structural role in cell walls of two algae that contain hypro-protein. Nor does there appear to be any evidence for splitting of the polypeptide backbone of cell wall hypro-protein during growth; experiments to detect turnover of this component have had negative results (Olson, 1964; Cleland, 1968d).

Workers in the plant growth field have been preoccupied with the idea of covalent cross-links between polymer chains as the basis for the behavior of the cell wall as an apparent network structure, but it should be emphasized that network behavior may arise by the cooperative action of secondary forces, such as H bonds, between polymer chains. The classical example is gelatin, whose gels are cross-linked by local regions of virtual crystallization. We have observed that dilute solutions of the hemicellulosic glucan of oat coleoptile cell walls (Wada and Ray, 1963) will set to a gel upon cooling, and show hydrodynamic properties indicative of extensive interactions between polymer chains, even though they contain no acidic groups and are patently not covalently cross-linked. In the condensed structure of the cell wall such secondary interactions may well dominate the rheological properties.

Wall Polymer Chain Degradation

Two principal lines of evidence suggest that breakdown of wall polymers may be involved in wall expansion and the action of auxin thereon. First, there are reports that breakdown or turnover of wall polysaccharides occurs. This has been detected by chemical analysis in some instances in which the total amount of a wall constituent is found to decrease during growth (Ray, 1963; Nelmes and Preston, 1968; and others). One kind of radioisotope evidence that has been given for turnover is that during feeding with labeled sugar, isotope incorporation into the cell wall can be greater than can be accounted for by the increase in amounts of wall components, suggesting that breakdown of preexisting unlabeled material must be occurring

(Maclachlan and Duda, 1965; Katz and Ordin, 1967a). This evidence is so dependent upon completeness of recovery and upon accuracy of chemical or gravimetric determinations that I do not find it conclusive. A second kind of isotope evidence for turnover is decrease in amount of label in particular wall constituents during growth in unlabeled medium following a pulse. Such results suggest at most a slow turnover of some noncellulosic wall fractions, especially glucan (Matchett and Nance, 1962; Baker and Ray, 1965a; Katz and Ordin, 1967a; Wada *et al.*, 1968; Roberts and Butt, 1969).

The published evidence on wall polysaccharide turnover has major defects and must be regarded as weak. First, starch is present in normal cell wall preparations and it has not been shown that decreases in, or chasing of label from, noncellulosic glucan is not due to depletion of starch. Second, decline in label has been detected only in certain extractable fractions while label was rising in other fractions, generally the less easily soluble ones. The decrease in label could have been due to a change in extractability of the labeled polymers, due to causes such as continued wall deposition, changes in physical association of the polymers with time, or modifications such as side-chain or ester-group addition or removal that would not involve any proper turnover of the polymer backbone. None of these possibilities has been distinguished conclusively. Perhaps the greatest weakness of wall turnover studies to date is that they have not dealt with individual types of macromolecules that comprise the cell wall.

A further difficulty with wall turnover studies as they relate to growth becomes apparent from cytological observations. Gross breakdown of the matrix (nonfibrillar) part of the primary wall occurs in the final stage of differentiation of xylem elements (Cronshaw and Bouck, 1965; O'Brien and Thimann, 1967; Pate *et al.*, 1969). This is illustrated in Fig. 8, which shows how the matrix material of primary walls of oat coleoptile xylem cells disappears except where near to or covered by secondary thickenings. A similar phenomenon can be seen in vessels of pea stems but appears not to have been recorded in the literature. It is a remarkably localized process, extending just to the middle lamella separating the wall of a xylem cell from that of the adjacent cell. One also notices a breakdown of wall material in the degenerating protophloem sieve tubes of oat coleoptiles.

Matrix breakdown may be important in permitting continued extension of the walls of these dead xylem and phloem cells and in this sense contributes to growth of the tissue. Since differentiation of xylem cells (and degeneration of sieve tubes) is going on during the usual growth and wall turnover experiments, it follows that any biochemically observed polymer breakdown might be involved entirely in these differentiation processes rather than in the growth of the living cells of the tissue.

Some attempts to detect promotion of wall polysaccharide turnover by auxin have given negative or inconclusive results (Matchett and Nance, 1962; Katz and Ordin, 1967a), but one instance of a positive effect has been claimed (Wada *et al.*, 1968). In view of the weaknesses of the methods used thus far, it seems possible that significant turnover effects remain to be established. On the other hand, since auxin promotes vascular cell differentiation, the involvement of any such effect in the cell enlargement process must be viewed with skepticism until it is established that it is not due to wall breakdown in differentiating cells.

It may also be remarked that isolated single cleavages of polymer chains, which would have a profound effect upon strain in a condensed or network structure, could occur without releasing any label from the polymers and thus without any turnover in the chemical sense. This would be detectable in terms of molecular weights of wall macromolecules but seems not to have been looked for.

A second line of evidence regarding wall degradation in relation to growth comes from observations on polysaccharidases. Several authors have reported detecting β-glucanases in growing plant tissues, in some cases associated with wall material (Lee *et al.*, 1967; Katz and Ordin, 1967b). These enzymes have not been completely characterized but include both cellulase (β-1,4-glucanase) (Maclachlan and Perrault, 1964; Konar and Stanley, 1969) and β-1,3- and β-1,6-glucanases that attack hemicellulosic glucans and callose (Clarke and Stone, 1962; Gamborg and Eveleigh, 1968; Wessels, 1969).

FIGS. 8 and 9. Autoradiographs of transverse sections of cells from elongating *Avena* coleoptile segments fed glucose-^{3}H. The opaque, round, or comma-shaped granules are silver grains.

FIG. 8. Xylem cells. Heavy incorporation of radioactivity into secondary thickenings of differentiating cells can be seen, and breakdown of the primary wall is evident (at arrows) in vessel element whose differentiation is just being completed by degeneration of protoplast. $\times$ 2800.

FIG. 9. Outer epidermis, showing incorporation throughout the thickness of the wall, but most heavily at inner surface. $\times$ 3750.

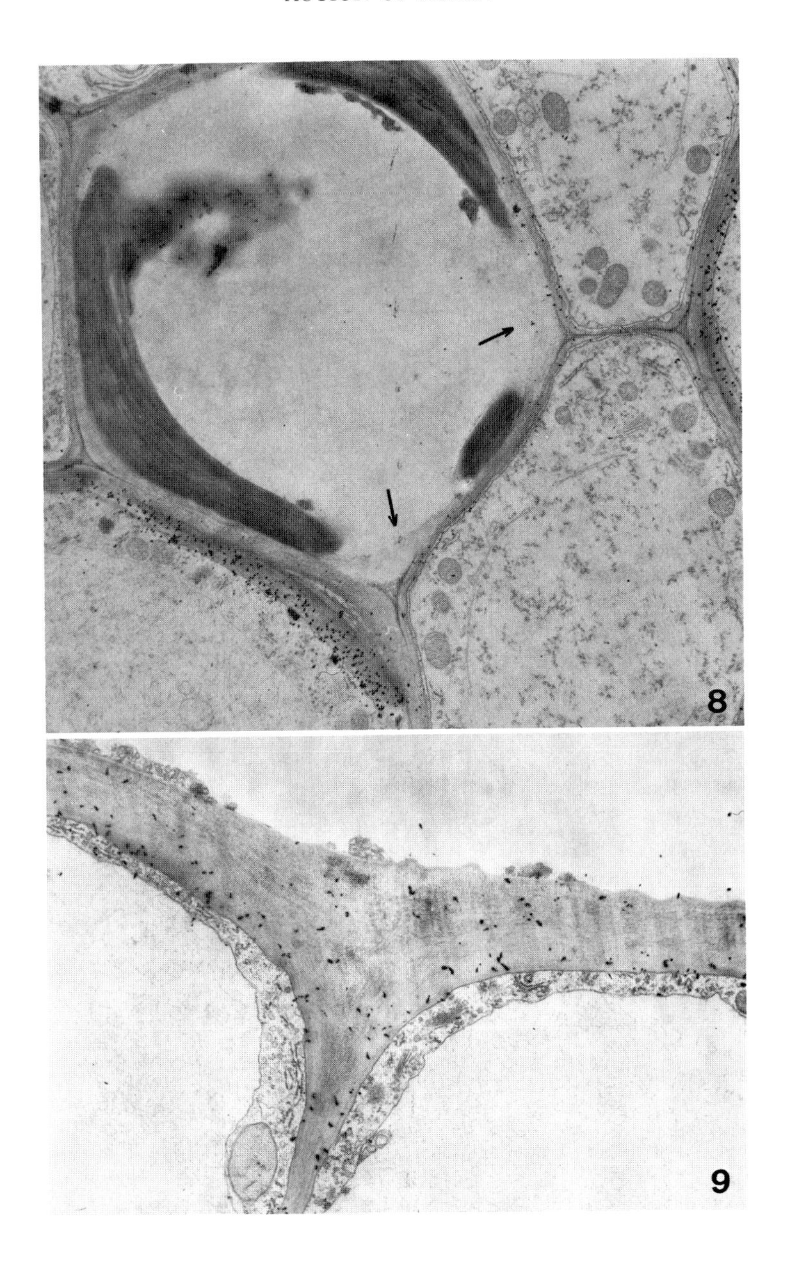
8
9

Fan and Maclachlan (1966, 1967a, b) found that massive increase in cellulase activity is induced by auxin in pea stem tissue; Davies and Maclachlan (1969) reported that this effect could be detected in cellulase synthesis by a microsomal preparation derived from auxin-treated tissue. The auxin effect on cellulase is slow (days), is found only under treatment with a high level of auxin in the range that inhibits elongation and promotes lateral swelling of the tissue, and seems to have nothing to do with the normal elongation growth of pea cells at physiological levels of auxin. It could lead to the lateral swelling response by weakening the oriented microfibrillar reinforcement upon which elongation depends.

Tanimoto and Masuda (1968) and Katz and Ordin (1967b) reported that treatment of oat coleoptile and pea stem segments with auxin at elongation-promoting levels caused increase in extractable hydrolase activity measured on hemicelluloses. Masuda and Wada (1967) have found that treatment of oat coleoptile segments with a particular fungal β-1,3-glucanase induces a rapid cell elongation and an increase in wall extensibility (Masuda, 1968) comparable to the effect of auxin itself, except that the effect lasts for only a short time. They propose that the action of auxin on growth is to induce synthesis of β-1,3-glucanase which is released into the cell wall and increases its extensibility by degrading hemicellulosic glucan.

Davies and Maclachlan (1968) and Datko and Maclachlan (1968) found no specific effect of auxin on β-1,3-glucanase activity of pea tissue. Cleland (1968c) found no promotion of coleoptile elongation by a purified fungal β-1,3-glucanase, and Ruesink (1968) similarly found no promotion by a fungal glucanase that was capable of releasing healthy protoplasts and that caused a demonstrable breakdown of wall material during the incubation. Some doubt about the nature of the effect found by Masuda and Wada (1967) is raised by their observation that heated enzyme also induced elongation, but to a lesser extent and after a relatively long latent period, as if the heated preparation contained a low concentration of auxin. One wonders whether the effect of the unheated enzyme preparation on elongation may, in fact, have been due to auxin, some of which was destroyed on heating; the timing shown for action of the unheated enzyme preparation on elongation is remarkably similar to that of auxin itself.

The glucanase theory of wall growth and auxin action seems to conflict with the dynamics of the growth process and auxin response. The plant glucanases that have been reported are stable enzymes since their hydrolytic action continues steadily for many hours and is normally assayed after periods of hours. As discussed above, the timing of the auxin response shows that it cannot be due to promoted synthesis of a stable enzyme, and the rapid decline in elongation upon withdrawal of auxin or treatment with an auxin antagonist shows that the products of auxin that effect growth rate are short-lived. One cannot get around this by supposing that the glucanases leak rapidly from the cell wall space into the medium, because the described β-glucanases are bound tightly to the cell wall (Heyn, 1969).

Morrè and Eisinger (1968) claimed to have detected a small auxin-promoted increase in wall extensibility in cell walls of killed maize coleoptiles caused by (presumably enzymatic) factors released by an adjacent, growing coleoptile segment.

Cleland (1968c) found that treatment with cyanide caused a reversal of auxin-induced increase in wall extensibility; since hydrolase action is irreversible, he concluded that auxin action cannot be due to glucanases.

While the possibility has not been excluded that the reported auxin effect on glucanase activity is actually a reflection of differentiation phenomena as discussed above, some role of hydrolases in the wall expansion process seems probable and the popularity of this idea appears to be rising rapidly, despite the somewhat conflicting pieces of evidence. Roggen and Stanley (1969) reported promotion of pollen germination by β-1,3-glucanase and of pollen-tube elongation by β-1,4-glucanase and pectinase, suggesting a role of polysaccharidases in normal growth. A similar view was advanced by Sato (1968). Evidence has been adduced for involvement of polysaccharidases and cell wall degradation in wall expansion growth in several fungi (Wessels, 1966; Johnson, 1968; Thomas and Mullins, 1969) and *Acetabularia* (Werz, 1966).

Synthesis of Wall Polymers

Synthesis of wall material normally accompanies growth, and promotion by auxin of the synthesis of wall polysaccharides (Fig. 10) has been found wherever carefully looked for in tissues the growth of which is promotable by auxin. Promotion of wall hypro-protein synthesis has also been detected (Kuraishi *et al.*, 1967; Cleland,

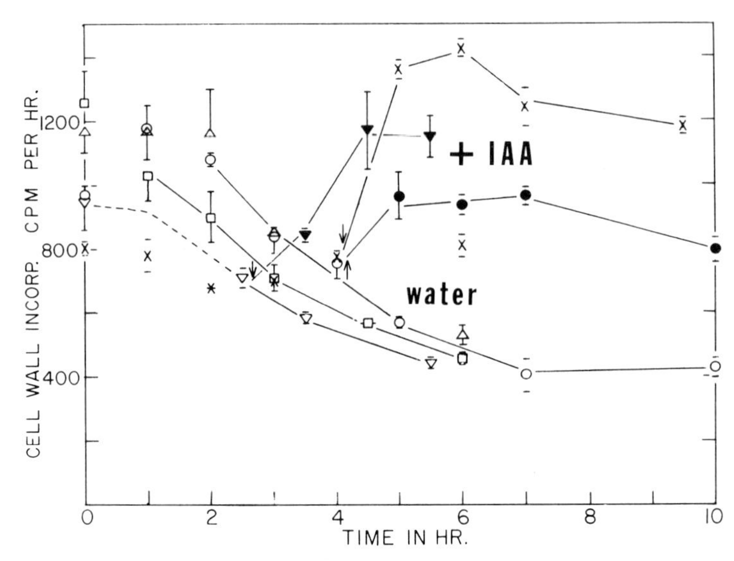

FIG. 10. Effect of auxin on cell wall synthesis by *Avena* coleoptile segments. Points show incorporation into cell walls during 1 hour incubations in 0.05 M ^{14}C-labeled glucose begun at the indicated times after cutting of segments, which were floated on water at 25°C prior to the isotope incubation and, in the 3 curves at upper right, had been transferred to 3 μg/ml IAA at times shown by arrows. Crosses, circles, triangles, and squares represent different experiments. Brackets show average deviation for replicate samples in each treatment.

1968b; Giesen and Klämbt, 1969). The promotion of wall polysaccharide synthesis by auxin is not due merely to the occurrence of elongation since it can be observed when elongation is inhibited by Ca^{2+} (Baker and Ray, 1965a; Ray and Abdul-Baki, 1968). The synthesis of all classes of wall matrix polysaccharides is promoted by auxin (Ray and Baker, 1965).

There is a strong correlation between promotive and inhibitory effects on wall synthesis and on growth rate that suggests a causative connection between these (Ray, 1962; Baker and Ray, 1965b). However, certain facts have repeatedly been cited as a basis for dismissing a causal role for wall biosynthesis in cell enlargement. Mannitol, at concentrations that inhibit growth osmotically, prevents the effect of auxin on wall synthesis although the effect of auxin on wall extensibility is still detectable (Cleland, 1967a). When tissues are treated with auxin in the absence of a substantial exogenous supply of sugar, the auxin effect on wall polysaccharide or protein synthesis is weak even though elongation is strongly promoted. During promotion of elongation by auxin, or inhibition of elongation by various agencies, e.g., light (Steiner, 1968, 1969), wall synthesis and elongation do not vary in close proportion, so that (for example) during the most rapid elongation the cell walls apparently get thinner whereas during repressed elongation they get thicker because of continued synthesis.

This last type of observation merely means that there must exist one type of synthesis which is *not* effective in inducing wall expansion and which continues even when expansion is not occurring. Once the possibility is recognized that two types of wall synthesis may exist, one involved in wall expansion (extensile synthesis) and the other not (intensile synthesis), it becomes obvious that these may compete with one another. When synthesis is restricted—for example, by shortage of substrate or by repressive conditions such as mannitol—one may anticipate that an acceleration of extensile synthesis will lead to a corresponding reduction of intensile synthesis, and vice versa, with little net effect on overall synthesis. Moreover, it was shown that the auxin effect on wall extensibility occurs much more slowly under mannitol inhibition (Cleland, 1967b). Therefore, the kinds of evidence cited above do not disprove an involvement of wall synthesis in cell growth.

Wall synthesis could be involved in stress relaxation and thus in wall expansion only if new material is introduced into and interacts with the existing wall structure which supports the turgor stress. Electron microscope autoradiography (Ray, 1967) showed that in growing epidermal cells of oat coleoptile and pea stem tissue new wall material is introduced throughout the thickness of the wall; however, a considerable fraction of the new material, including all the newly made cellulose, is deposited appositionally at the wall surface (Fig. 9). This demonstrates both the possibility of participation of wall synthesis in wall expansion and the existence of a component of total synthesis that cannot contribute to expansion.

The fraction of the total new wall material that is incorporated internally was found to be increased by auxin, and is large enough (approximately 50% in the case of coleoptiles) that it would be difficult to imagine that it did *not* contribute to wall expansion. Cell walls that do not undergo expansion growth—namely, secondary walls of the stomatal guard cells and of xylem elements—were observed to be synthesized by apposition.

Two modes of internal or intussusceptional incorporation may be recognized: (1) introduction of new subunits into the interior of a polymer molecule, as occurs in bacterial wall peptidoglycan synthesis (Izaki *et al.*, 1968), and which depends upon a concurrent splitting of the polymer chain (Schwartz *et al.*, 1969); (2) introduction of new macromolecules between existing ones in the wall structure, a process that would require exchange of bonds or bonding forces (ionic, hydrogen, and/or van der Waals) between preexisting macromolecules and newly introduced ones. The first mode has been inferred as the mechanism of wall growth in yeast (Johnson, 1968), because autolysis of wall glucan can be detected in the growing region of the wall when glucan synthesis is inhibited. However, hydrolase action could equally well contribute to macromolecular intussusception (mode 2) by splitting network chains and/or by opening up local holes that allow new polymer molecules to penetrate the structure and to interact with preexisting ones. There is no evidence for subunit intussusception within wall polymers of higher plants. The known polysaccharide synthetases that use sugar nucleotides as glycosyl donors are borne by cytoplasmic organelles (Ray *et al.*, 1969) and are not found in the cell wall itself. Hypro-protein synthesis is apparently ribosomal, as would be expected. Some further polymerization of preformed polymers within the cell wall seems possible, and could be detected by density-label experiments, but such experiments have not yet been performed.

Maximum promotion by auxin of synthesis of wall polysaccharides (Ray and Abdul-Baki, 1968, and unpublished data) and hypro-protein (Kuraishi *et al.*, 1968; Giesen and Klämbt, 1969) develops only an hour or more after exposure to auxin (Fig. 10) and thus is much slower than the elongation response. This shows that the promotion itself is not the cause of the elongation response. In this respect the auxin effect on wall synthesis is in about the same position relative to the growth response as the auxin effects on RNA and protein synthesis discussed earlier. However, it is easily possible that the net promotion of wall synthesis is a secondary regulatory response resulting from an initial diversion from intensile to extensile synthesis.

The regulation of wall polysaccharide synthesis has been explored in both pea stem (Ray and Abdul-Baki, 1968) and oat coleoptile (Hall and Ordin, 1968) tissue. The response of wall synthesis to auxin is reduced by inhibitors of protein synthesis. The only relevant enzyme the activity of which changes detectably in response to auxin treatment of the tissue is β-glucan synthetase (other polysaccharide synthetases may also change but have not yet been studied). The auxin effect on synthetase is abolished by cycloheximide (Hall and Ordin, 1968). The auxin effects on synthetase that have been published are rather small compared to the effect of auxin on the rate of *in vivo* wall synthesis, leading to doubt as to whether the synthetase response can account for the biological response. However, in more recent experiments we have obtained a much larger (2-fold) auxin effect on synthetase activity using an improved method of synthetase preparation and a somewhat longer period of exposure to auxin (Table 1). We have recently succeeded in isolating the synthetase-bearing particles (Ray *et al.*, 1969); they have proved to be Golgi membranes. This will make it possible to investigate in much more detail the nature of the auxin effect on synthetase activity which, in view of the membrane-bound character of the system, might involve regulation by compartmentation as well as by effectors (Camargo *et al.*, 1967; Thomas *et al.*, 1969; and references there cited) or by changes in enzyme content.

Cleland (1967d) found that inhibition of growth by free hydroxyproline could be correlated with inhibition of formation or hydroxylation of cell wall hypro-protein. This suggests a requirement for hypro-protein synthesis in growth. Cleland (1968b) showed that promotion by auxin of incorporation of labeled proline into hypro-protein of coleoptiles occurs only when sugar is supplied, i.e., under conditions that make possible a substantial wall polysaccharide syn-

TABLE 1
AUXIN EFFECT ON GLUCAN SYNTHETASE ACTIVITY OBTAINABLE FROM PEA TISSUE[a, b]

Pretreatment time (hours)	Tissue pretreatment	
	minus IAA	plus IAA
0	39.3	—
2	37.9	44.1
4	13.8	25.5

[a] Each figure is the average of assays on duplicate samples of 100 pea stem segments (1.9 gm initial fresh weight) that had been incubated in 0.1 *M* glucose with or without 3 μg of indoleacetic acid per milliliter for the indicated times, then homogenized and assayed for total particulate synthetase activity as described by Ray *et al.* (1969), using UDPG.

[b] Values are expressed as total activity in nanomoles per hour per 100 segments.

thesis response to auxin. He also found (Cleland, 1967c) that free hydroxyproline inhibited elongation only under conditions of sugar supply. He drew the conclusion that hypro-protein might be required to bring about internal incorporation of polysaccharides (Cleland, 1968b). Hypro-protein could function in this way by serving as a terminal glycosyl acceptor for polysaccharide chains that otherwise become attached to existing wall polysaccharides immediately upon exit through the plasma membrane.

Internal incorporation of hypro-protein into the cell wall seems not to have been directly demonstrated as yet. So far, hypro-protein is known to be an acceptor only for galactoarabinan chains (Lamport, 1967, 1969), whereas most of the material that becomes incorporated internally into oat coleoptile cell walls is hemicellulosic glucan and glucuronoarabinoxylan (Ray, 1967). Holleman (1967) reported that hydroxylation of protein-bound proline to hydroxyproline could be completely inhibited by α,α'-dipyridyl while only partially inhibiting the growth of *Acer pseudoplatanus* callus. This might suggest that hypro-protein is not required for growth, and Holleman's results indicated that direct incorporation of hydroxyproline into proteins might be the cause of its inhibitory effect on growth.

Despite these questions the idea of a glycoprotein-dependent intussusception mechanism seems very attractive since it might account for the evidence that both hypro-protein and polysaccharide synthesis plays a role in growth.

CONCLUSION

There is increasing evidence that both degradation of wall polymers and biosynthesis of new wall material participate in cell wall expansion and in the action of auxin thereon. I suggest that degradation and biosynthesis act cooperatively in the growth process; degradation both directly causes stress relaxation and also opens up sites for intussusceptive incorporation of new macromolecules (polysaccharide and glycoprotein) into the cell wall, while incorporation leads, by processes of weak bond interchange between preexisting wall polymers and newly introduced ones, to stress relaxation and therefore to strain. Ionic interactions between acidic polysaccharides, involving Ca^{2+}, appear to contribute to the plastic-yielding resistance of wall material, but at present it appears that these bonds yield passively during growth. Although the cell wall behaves as a condensed structure, there is as yet no evidence for a network of covalent cross-links between wall polymers nor for involvement of special cross-link-splitting processes in plant cell growth. The strain process visualized above differs basically from that indicated for bacterial cell walls in that it involves the introduction of new macromolecules, rather than internal extension of polymers by splitting and introduction of new subunits.

It is tempting to think that auxin acts on cell enlargement by inducing the formation of hydrolases and synthetases involved in the degradative and synthetic aspects of wall expansion; evidence exists for an auxin-induced increase in both hydrolase and synthetase activity. However, the timing of growth responses to auxin discourages acceptance of a gene activation mechanism of auxin action, and it seems more likely that the immediate effect of auxin on elongation is the result of stimulated transport of hydrolases and/or of new wall polymers or precursors thereof into the cell wall, with the observed auxin stimulation of enzyme activity and net biosynthetic rate being consequences of regulatory mechanisms that are coupled to the utilization of these materials in the growth process.

ACKNOWLEDGMENTS

The original research reviewed here was supported by grants from the National Science Foundation. I thank Dr. Michael Evans for making available unpublished results, and Dr. Paul Green and Dr. Robert Cleland for valuable discussions.

REFERENCES

Andreae, W. A., and Collet, G. (1968). The effect of phenolic substances on the growth activity of indoleacetic acid applied to pea root or stem sections. *In* "Biochemistry and Physiology of Plant Growth Substances" (F. Wightman and G. Setterfield, eds.), pp. 553–561. Runge Press, Ottawa.

Baker, D. B., and Ray, P. M. (1965a). Direct and indirect effects of auxin on cell wall synthesis in oat coleoptile tissue. *Plant Physiol.* **40,** 345–352.

Baker, D. B., and Ray, P. M. (1965b). Relationship between effects of auxin on cell wall synthesis and cell elongation. *Plant Physiol.* **40,** 360–368.

Barkley, G. M., and Evans, M. L. (1969). Timing of the auxin response in etiolated pea stem sections. *Plant Physiol.* (in press).

Black, M., Bullock, C., Chantler, E. N., Clarke, R. A., Hanson, A. D., and Jolley, G. M. (1967). Effect of inhibitors of protein synthesis on the plastic deformation and growth of plant tissues. *Nature* **215,** 1289–1290.

Bottrill, D. E., and Hanson, J. B. (1968). Short-term growth response to growth regulators in roots of *Zea mays*. *Australian J. Biol. Sci.* **21,** 201–208.

Brian, R. C., and Rideal, E. K. (1952). On the action of plant growth regulators. *Biochim. Biophys. Acta* **9,** 1–18.

Burg, S. P., and Burg, E. A. (1968). Auxin stimulated ethylene formation: its relationship to auxin inhibited growth, root geotropism and other plant processes. *In* "Biochemistry and Physiology of Plant Growth Substances" (F. Wightman and G. Setterfield, eds.), pp. 1275–1294. Runge Press, Ottawa.

Camargo, E. P., Dietrick, C. P., Sonneborn, D., and Strominger, J. L. (1967). Biosynthesis of chitin in spores and growing cells of *Blastocladiella emersonii*. *J. Biol. Chem.* **242,** 3121–3128.

Clarke, A. E., and Stone, B. A. (1962). β-1,3-glucan hydrolases from the grape vine (*Vitis vinifera*) and other plants. *Phytochemistry* **1,** 175–188.

Cleland, R. (1959). Effect of osmotic concentration on auxin action and on irreversible and reversible expansion of the Avena coleoptile. *Physiol. Plantarum* **12,** 809–825.

Cleland, R. (1960). Effect of auxin upon loss of calcium from cell walls. *Plant Physiol.* **35,** 581–584.

Cleland, R. (1963a). Independence of effects of auxin on cell wall methylation and elongation. *Plant Physiol.* **38,** 12–18.

Cleland, R. (1963b). The occurrence of auxin-induced pectin methylation in plant tissues. *Plant Physiol.* **38,** 738–740.

Cleland, R. (1965). Auxin-induced cell wall loosening in the presence of actinomycin D. *Plant Physiol.* **40,** 595–600.

Cleland, R. (1967a). Extensibility of isolated cell walls: measurement and changes during cell elongation. *Planta* **74,** 197–209.

Cleland, R. (1967b). A dual role of turgor pressure in auxin-induced cell elongation in Avena coleoptiles. *Planta* **77,** 182–191.

Cleland, R. (1967c). Inhibition of cell elongation in Avena coleoptile by hydroxyproline. *Plant Physiol.* **42,** 271–274.

Cleland, R. (1967d). Inhibition of formation of protein-bound hydroxyproline by free hydroxyproline in Avena coleoptiles. *Plant Physiol.* **42,** 1165–1170.

Cleland, R. (1968a). Wall extensibility and the mechanism of auxin-induced cell elongation. *In* "Biochemistry and Physiology of Plant Growth Substances" (F. Wightman and G. Setterfield, eds.), pp. 613–624. Runge Press, Ottawa.

Cleland, R. (1968b). Hydroxyproline formation and its relation to auxin-induced cell elongation in the Avena coleoptile. *Plant Physiol.* **43,** 1625–1630.

Cleland, R. (1968c). Auxin and wall extensibility: reversibility of auxin-induced wall-loosening process. *Science* **160,** 192–193.

Cleland, R. (1968d). The distribution and metabolism of protein-bound hydroxyproline in an elongating tissue, the Avena coleoptile. *Plant Physiol.* **43,** 865–870.

Cronshaw, J., and Bouck, G. B. (1965). The fine structure of differentiating xylem elements. *J. Cell. Biol.* **24,** 415–431.

Datko, A. H., and Maclachlan, G. A. (1968). IAA and the synthesis of glucanases and pectic enzymes. *Plant Physiol.* **43,** 735–742.

Davies, E., and Maclachlan, G. A. (1968). Effects of indoleacetic acid on intracellular distribution of β-glucanase activities in the pea epicotyl. *Arch. Biochem. Biophys.* **128,** 595–600.

Davies, E., and Maclachlan, G. A. (1969). Generation of cellulase activity during protein synthesis by pea microsomes *in vitro*. *Arch. Biochem. Biophys.* **129,** 581–587.

Evans, M. L. (1967). Kinetic studies of the cell elongation phenomenon in *Avena* coleoptile segments. Ph.D. Thesis, Univ. of California, Santa Cruz.

EVANS, M. L., and HOKANSON, R. (1969). Timing of the response of coleoptiles to the application and withdrawal of various auxins. *Planta* **85,** 85–95.

EVANS, M. L., and RAY, P. M. (1969). Timing of the auxin response in coleoptiles and its implications regarding auxin action. *J. Gen. Physiol.* **53,** 1–20.

FAN, D. F., and MACLACHLAN, G. A. (1966). Control of cellulase activity by IAA. *Can. J. Botany* **44,** 1025–1034.

FAN, D. F., and MACLACHLAN, G. A. (1967a). Massive synthesis of RNA and cellulase in response to IAA. *Plant Physiol.* **42,** 1114–1122.

FAN, D. F., and MACLACHLAN, G. A. (1967b). Studies on the regulation of cellulase activities and growth in excised pea epicotyl sections. *Can. J. Botany* **45,** 1837–1844.

GAMBORG, O. L., and EVELEIGH, D. E. (1968). Culture methods and detection of glucanases in suspension cultures of wheat and barley. *Can. J. Biochem.* **46,** 417–421.

GIESEN, M., and KLÄMBT, D. (1969). Untersuchungen zur Beeinflussung des Prolin-Einbaues in Proteine durch Auxin und Hydroxyprolin in Weizenkoleoptilcylindern. *Planta* **85,** 73–84.

GREEN, P. B. (1968). Growth physics in Nitella: a method for continuous *in vivo* analysis of extensibility based on a micro-manometer technique for turgor pressure. *Plant Physiol.* **43,** 1169–1184.

HALL, M. A., and ORDIN, L. (1968). Auxin-induced control of cellulose synthetase activity in *Avena* coleoptile sections. *In* "Biochemistry and Physiology of Plant Growth Substances" (F. Wightman and G. Setterfield, eds.), pp. 659–675. Runge Press, Ottawa.

HASCHEMEYER, A. E. V. (1969). Rates of polypeptide chain assembly in liver *in vivo*: relation to the mechanism of temperature acclimation in *Opsanus tau* (the toad fish). *Proc. Natl. Acad. Sci.* **62,** 128–135.

HEJNOWICZ, Z., and ERICKSON, R. O. (1968). Growth inhibition and recovery in roots following temporary treatment with auxin. *Physiol. Plantarum* **21,** 302–313.

HERTEL, R., EVANS, M. L., LEOPOLD, A. C., and SELL, H. M. (1969). The specificity of the auxin transport system. *Planta* **85,** 238–249.

HERTEL, R., and FLORY, R. (1968). Auxin movement in corn coleoptiles. *Planta* **82,** 123–144.

HEYN, A. N. J. (1969). Glucanase activity in coleoptiles of Avena. *Arch. Biochem. Biophys.* **132,** 442–449.

HOLLEMAN, J. (1967). Direct incorporation of hydroxyproline into protein of sycamore cells incubated at growth-inhibitory levels of hydroxyproline. *Proc. Natl. Acad. Sci. U.S.* **57,** 50–54.

HUBBELL, W. L., and MCCONNELL, H. L. (1968). Spin-label studies on the excitable membranes of nerve and muscle. *Proc. Natl. Acad. Sci. U.S.* **61,** 12–16.

HUBBELL, W. L., and MCCONNELL, H. L. (1969). Motion of steroid spin labels in membranes. *Proc. Natl. Acad. Sci. U.S.* **63,** 16–27.

IZAKI, K., MATSUHASHI, M., and STROMINGER, J. L. (1968). Biosynthesis of the peptidoglycan of bacterial cell walls. XIII. Peptidoglycan transpeptidase and D-alanine carboxypeptidase: penicillin-sensitive enzymatic reactions in strains of *Escherichia coli*. *J. Biol. Chem.* **243,** 3180–3192.

JOHNSON, B. F. (1968). Dissolution of yeast glucan induced by 2-deoxyglucose. *Exptl. Cell. Res.* **50,** 692–694.

KATZ, M., and ORDIN, L. (1967a). Metabolic turnover in cell wall constituents of *Avena sativa* L. coleoptile sections. *Biochim. Biophys. Acta* **141,** 118–125.

KATZ, M., and ORDIN, L. (1967b). A cell wall polysaccharide-hydrolyzing enzyme system in *Avena sativa* L. coleoptiles. *Biochim. Biophys. Acta* **141,** 126–134.

KAUSS, H., and SWANSON, A. L. (1969). Cooperation of enzymes responsible for polymerization and methylation in pectin biosynthesis. *Z. Naturforsch.* **24b,** 28–33.

KEY, J. L. (1969). Hormones and nucleic acid metabolism. *Ann. Rev. Plant Physiol.* **20,** 449–474.

KONAR, R. N., and STANLEY, R. G. (1969). Wall-softening enzymes in the gynoecium and pollen of *Hemerocallis fulva*. *Planta* **84,** 304–310.

KURAISHI, S., UEMATSU, S., and YAMAKI, T. (1967). Auxin-induced incorporation of proline in mung bean hypocotyls. *Plant Cell Physiol.* **8,** 527–528.

KURAISHI, S., KASAMO, K., and YAMAKI, T. (1968). The relationship between growth and proline incorporation after auxin treatment of mung bean hypocotyls. *Plant Cell Physiol.* **9,** 842–850.

LAMPORT, D. T. A. (1965). The protein component of primary cell walls. *Advan. Botan. Res.* **2,** 151–218.

LAMPORT, D. T. A. (1967). Hydroxyproline-*O*-glycosidic linkage of the plant cell wall glycoprotein extensin. *Nature* **216,** 1322.

LAMPORT, D. T. A. (1969). The isolation and partial characterization of hydroxyproline-rich glycopeptides obtained by enzymic degradation of primary cell walls. *Biochemistry* **8,** 1155–1163.

LEE, S., KIVILAAN, A., and BANDURSKI, R. S. (1967). *In vitro* autolysis of plant cell walls. *Plant Physiol.* **42,** 968–972.

LOCKHART, J. A. (1965). Cell extension. *In* "Plant Biochemistry" (J. Bonner and J. E. Varner, eds.), pp. 826–849. Academic Press, New York.

LOCKHART, J. A. (1967). Physical nature of irreversible deformation of plant cells. *Plant Physiol.* **42,** 1545–1552.

MACDONALD, I. R., and ELLIS, R. J. (1969). Does cycloheximide inhibit protein synthesis specifically in plant tissues? *Nature* **222,** 791–792.

MACLACHLAN, G. A., and DUDA, C. (1965). Changes in concentration of polymeric components in excised pea-epicotyl tissue during growth. *Biochim. Biophys. Acta* **97,** 288–299.

MACLACHLAN, G. A., and PERRAULT, J. (1964). Cellulase from pea epicotyls. *Nature* **204,** 81–82.

MASUDA, Y. (1968). Role of cell-wall-degrading enzymes in cell-wall loosening in oat coleoptiles. *Planta* **83,** 171–184.

MASUDA, Y. (1969). Auxin-induced cell expansion in relation to cell wall extensibility. *Plant Cell Physiol.* **10,** 1–10.

MASUDA, Y., and KAMISAKA, S. (1969). Rapid stimulation of RNA biosynthesis by auxin. *Plant Cell Physiol.* **10,** 79–86.

MASUDA, Y., and WADA, S. (1967). Effect of β-1,3-glucanase on the elongation growth of oat coleoptile. *Botan. Mag.* (*Tokyo*) **80,** 100–102.

MATCHETT, W. H., and NANCE, J. F. (1962). Cell wall breakdown and growth in pea seedling stems. *Am. J. Botany* **49,** 311–319.

MATTHYSSE, A. (1968). The effect of auxin on RNA synthesis. *Plant Physiol.* **43,** suppl, S-42 (Abstract).

MITCHELL, K. K., and SARKISSIAN, I. V. (1966). Effects of indol-3-yl acetic acid on the citrate-condensing reaction by preparations of root and shoot tissue. *J. Exptl. Botany* **17,** 838–843.

MORRÉ, D. J., and EISINGER, W. R. (1968). Cell wall extensibility: its control by auxin and relationship to cell elongation. *In* "Biochemistry and Physiology of Plant Growth Substances" (F. Wightman and G. Setterfield, eds.), pp. 625–645. Runge Press, Ottawa.

MOYED, H. S., and TULI, V. (1968). The oxindole pathway of 3-indoleacetic acid metabolism and the action of auxins. *In* "Biochemistry and Physiology of Plant Growth Substances" (F. Wightman and G. Setterfield, eds.), pp. 289–300. Runge Press, Ottawa.

NELMES, B. J., and PRESTON, R. D. (1968). Wall development in apple fruits: a study of the life history of a parenchyma cell. *J. Exptl. Botany* **19,** 496–518.

NICKERSON, W. J. (1963). Molecular bases of form in yeasts. *Bacteriol. Rev.* **27,** 305–324.

O'BRIEN, T. P., and THIMANN, K. V. (1967). Observations on the fine structure of the oat coleoptile. III. Correlated light and electron microscopy of the vascular tissues. *Protoplasma* **63,** 443–478.

OLSON, A. C. (1964). Proteins and plant cells walls. Proline to hydroxyproline in tobacco suspension cultures. *Plant Physiol.* **39,** 543–550.

ORDIN, L., CLELAND, R., and BONNER, J. (1957). Methyl esterification of cell wall constituents under the influence of auxin. *Plant Physiol.* **32,** 216–220.

PATE, J. S., GUNNING, B. E. S., and BRIARTY, L. G. (1969). Ultrastructure and functioning of the transport system of the leguminous root nodule. *Planta* **85,** 11–34.

PATTERSON, B. D., and TREWAVAS, A. J. (1967). Changes in the pattern of protein synthesis induced by 3-indolylacetic acid. *Plant Physiol.* **42,** 1081–1086.

POOLE, R. J., and THIMANN, K. V. (1964). Uptake of indole-3-acetic acid and indole-3-acetonitrile by *Avena* coleoptile sections. *Plant Physiol.* **39,** 98–103.

PRESTON, R. D., and HEPTON, J. (1960). The effect of indoleacetic acid on cell wall extensibility in *Avena* coleoptiles. *J. Exptl. Botany* **11,** 13–27.

RAY, P. M. (1961). Hormonal regulation of plant cell growth. *In* "Control Mechanisms in Cellular Processes" (D. M. Bonner, ed.), pp. 185–212. Ronald Press, New York.

RAY, P. M. (1962). Cell wall synthesis and cell elongation in oat coleoptile tissue. *Am. J. Botony* **49,** 928–939.

RAY, P. M. (1963). Sugar composition of oat-coleoptile cell walls. *Biochem. J.* **89,** 144–150.

RAY, P. M. (1967). Radioautographic study of cell wall deposition in growing plant cells. *J. Cell Biol.* **35,** 659–674.

RAY, P. M., and ABDUL-BAKI, A. (1968). Regulation of cell wall synthesis in response to auxin. *In* "Biochemistry and Physiology of Plant Growth Substances" (F. Wightman and G. Setterfield, ed.), pp. 647–658. Runge Press, Ottawa.

RAY, P. M., and BAKER, D. B. (1965). The effect of auxin on synthesis of oat coleoptile cell wall constituents. *Plant Physiol.* **40,** 353–360.

RAY, P. M., and RUESINK, A. W. (1962). Kinetic experiments on the nature of the growth mechanism in oat coleoptile cells. *Develop. Biol.* **4,** 377–397.

RAY, P. M., and RUESINK, A. W. (1963). Osmotic behavior of oat coleoptile tissue in relation to growth. *J. Gen. Physiol.* **47,** 83–101.

RAY, P. M., SHININGER, T. L., and RAY, M. M. (1969). Isolation of β-glucan synthetase particles from plant cells and identification with Golgi membranes. *Proc. Natl. Acad. Sci. U.S.* **63,** in press.

REINHOLD, L., and GLINKA, Z. (1966). Reduction in turgor pressure as a result of extremely brief exposure to CO_2. *Plant Physiol.* **41,** 39–44.

ROBERTS, R. M., and BUTT, V. S. (1969). Patterns of incorporation of D-galactose into cell wall polysaccharide of growing maize roots. *Planta* **84,** 250–262.

ROGGEN, H. P., and STANLEY, R. G. (1969). Cell-wall-hydrolysing enzymes in wall formation as measured by pollen-tube extension. *Planta* **84,** 295–303.

RUESINK, A. W. (1968). Fungal cellulase treatment of higher plant cell walls. *Plant Physiol.* **43,** Suppl, S-24 (Abstract).

SARKISSIAN, I. V., and SCHMALSTIEG, F. C. (1969). Confirmation of the effect of indole-3-acetic acid on citrate synthetase. *Naturwissenschaften* **56,** 284.

SATO, S. (1968). Enzymatic maceration of plant tissue. *Physiol. Plantarum* **21,** 1067–1075.

SCHWARZ, U., ASMUS, A., and FRANK, H. (1969). Autolytic enzymes and cell division of *Escherichia coli*. *J. Mol. Biol.* **41,** 419–429.

STEINER, A. M. (1968). Änderungen im Gehalt an löslichen Zuckern und Zellwandkohlenhydraten während der Hemmung endogenen Zellstreckungswachstums durch Antimetaboliten. *Planta* **83,** 282–294.

STEINER, A. M. (1969). Die Zellwandzusammensetzung von Hypokotylen des Senfkeimlings (*Sinapis alba* L.) bei Wachstumshemmung durch Inhibitoren oder Phytochrom. *Planta* **84,** 348–352.

STROMINGER, J. L., IZAKI, K., MATSUHASHI, M., and TIPPER, D. J. (1967). Peptidoglycan transpeptidase and D-alanine carboxypeptidase: penicillin-sensitive enzymatic reactions. *Federation Proc.* **26,** 9–22.

SWEENEY, B. M., and THIMANN, K. V. (1938). The effect of auxins on protoplasmic streaming. II. *J. Gen. Physiol.* **21,** 439–461.

TANIMOTO, E., and MASUDA, Y. (1968). Effect of auxin on cell wall degrading enzymes. *Physiol. Plantarum* **21,** 820–826.

THOMAS, D. S., and MULLINS, J. T. (1969). Cellulase induction and wall extension in the water mold *Achlya ambisexualis*. *Physiol. Plantarum* **22,** 347–353.

THOMAS, D. S., SMITH, J. E., and STANLEY, R. G. (1969). Stereospecific regulation of plant glucan synthetase *in vivo*. *Can. J. Botany* **47,** 489–496.

THOMPSON, E. W., and PRESTON, R. D. (1968). Evidence for a structural role of protein in algal cell walls. *J. Exptl. Botany* **19,** 690–697.

UHRSTRÖM, I. (1969). The time effect of auxin and calcium on growth and elastic modulus in hypocotyls. *Physiol. Plantarum* **22,** 271–287.

VAN OVERBEEK, J. (1959). Auxins. *Botan. Rev.* **25,** 269–350.

VAN OVERBEEK, J. (1961). New theory on the primary mode of auxin action. *In* "Plant Growth Regulation" (R. M. Klein, ed.), pp. 449–455. Iowa State Univ. Press, Ames, Iowa.

VELDSTRA, H. (1953). The relation of chemical structure to biological activity in growth substances. *Ann. Rev. Plant Physiol.* **4,** 151–198.

WADA, S., and RAY, P. M. (1963). Polysaccharides of the hemicellulose fraction of oat coleoptile cell walls. *Plant Physiol.* **38,** Suppl. xiii (Abstract).

WADA, S., TANIMOTO, E., and MASUDA, Y. (1968). Cell elongation and metabolic turnover of the cell wall as affected by auxin and cell wall degrading enzymes. *Plant Cell Physiol.* **9,** 369–376.

WEIGL, J. (1969). Einbau von Auxin in gequollene Lecithin-Lamellen. *Z. Naturforsch.* **24b,** 365–366.

WERZ, G. (1966). Primärvorgänge bei der Realisation der Morphogenese von *Acetabularia*. *Planta* **69**, 53–57.

WESSELS, J. G. H. (1966). Control of cell-wall glucan degradation during development in *Schizophyllum commune*. *Antonie van Leeuwenhoek, J. Microbiol. Serol.* **32**, 341–355.

WESSELS, J. G. H. (1969). A β-1,6-glucan glucanohydrolase involved in hydrolysis of cell wall glucan in *Schizophyllum commune*. *Biochim. Biophys. Acta* **178**, 191–193.

ZENK, M. H., and NISSL, D. (1968). Evidence against an allosteric effect of indole-3-acetic acid on citrate synthetase. *Naturwissenschaften* **55**, 84–85.

DEVELOPMENTAL BIOLOGY SUPPLEMENT 3, 206–226 (1969)

Control of Nutrient Assimilation, A Growth-Regulating Mechanism in Cultured Plant Cells

PHILIP FILNER

MSU/AEC Plant Research Laboratory, Michigan State University, East Lansing, Michigan 48823

METABOLITES AS DEVELOPMENTAL MESSAGES

Many of our present ideas about the nature of regulatory mechanisms at the molecular level have developed out of studies of regulation of enzyme formation and function in bacteria and other microorganisms (Jacob and Monod, 1963). The validity of many of the principles derived from the microbial studies has not yet been established for complex, multicellular organisms. In spite of this information gap, there has been a great deal of speculation about how mechanisms of the microbial type might generate the orderly sequences of events in multicellular organisms which we call development.

A key feature of the mechanisms which have been exhaustively studied in bacteria is that they appear to be designed to maintain production of an essential metabolite at the rate required to sustain the maximum possible growth rate in a given environment. That is, they regulate syntheses of metabolites, and they function to resist change in the growth rate of the organism. The metabolites themselves serve as the sources of information which govern these mechanisms.

The mechanisms of development in multicellular organisms appear to differ markedly. It is in their very nature to facilitate change rather than to resist it. The best-known agents for initiating developmental sequences, the hormones, seem to be arbitrary carriers of information; there is no relationship yet known between the metabolism of a hormone and the information it carries, as judged from the responses which hormones evoke. The latter feature of developmental mechanisms in higher organisms—the widespread lack of a metabolic relationship between the medium and the message—is unfortunate from the point of view of this investigator, since the existence of the relationship has been the key to much of the progress in understanding the bacterial mechanisms. The relationship, when it exists, focuses attention on the enzymes of the metabolism of the message, and the portions of the genome which specify those enzymes, as possible sites of analysis of and response to the message. If the message is arbitrarily assigned to a molecule, then the problem of finding the site of initiation of the biochemical chain of events, to say nothing of determining the biochemical events which account for the developmental physiology, becomes far more difficult.

Wouldn't it be nice to have systems in which a developmental event is initiated by a metabolite, particularly one whose biochemistry is well known? Then, with a little bit of luck, perhaps the site of initiation and the chain of biochemical steps leading to the developmental event would be found in the known biochemistry.

We do not customarily think of metabolites as initiators of developmental sequences, but there is no immutable law of Nature which says that these cannot function in this capacity. Metabolites should not be overlooked as possible developmental effectors.

An apparent example is the influence of the available nitrogen level on the relative development of roots and shoots of plants (Sinnott, 1960). In many species, the rate of development of the root system relative to the rate of development of the shoot increases as the nitrogen available through the roots decreases, while within a certain range the rate of development of the shoot is relatively insensitive to the level of nitrogen available. Apparently the development of the root system is accelerated or decelerated with respect to the development of the shoot until its capacity to scavenge nitrogen from the environment is attuned to the nitrogen demand of the shoot during optimal growth. Nitrogen in some form must serve as a message somewhere in this process which determines the gross form of the plant.

Other effects of nutrients, particularly nitrogen, on plant morphogenesis are known (Sinnott, 1960). Amino acids have striking effects on the morphogenesis of tobacco (Steinberg, 1949), peas (Fries, 1954), and *Oenanthe* (Waris, 1959, 1962). Other metabolites also alter the morphogenesis of certain plants (Allsopp, 1964).

THE NITRATE PATHWAY, A COMMUNICATION LINE

Plants normally obtain nitrogen as nitrate. Some plants, such as cocklebur, take up the nitrate through the cells of the roots but appear to translocate the nitrate without reducing it, and then reduce it in the stems and leaves (Wallace and Pate, 1967). In the case of cocklebur, we thus have an apparent example of the nitrate reducing pathway being absent in one part of the plant, the roots, and present in another, the stems and leaves. There is no intrinsic problem in reducing nitrogen in roots, however. The field pea apparently reduces nitrate in its roots and translocates reduced forms of nitrogen to the upper parts of the plant (Wallace and Pate, 1965). If given a high level of nitrate, the field pea will begin to translocate the excess nitrate to the upper parts and reduce it there. In any event, assimilation of nitrate involves coordination of activities of both the upper and lower parts of plants, and there are at least two satisfactory solutions to the problem. In one case, the upper part of the plant receives reduced nitrogen compounds, and in the other case, the upper part of the plant receives nitrate and reduces it.

The assimilatory nitrate pathway converts nitrate from the environment to ammonium via a system including a nitrate uptake mechanism and two enzymes, nitrate reductase and nitrite reductase (Fig. 1), thereby making the nitrogen available for a variety of biosyntheses, the most important one, quantitatively, being protein synthesis (Kessler, 1964; Bandurski, 1965; Beevers and Hageman, 1969). The pathway can be rendered superfluous and potentially harmful by the presence of any reduced nitrogen compound which can substitute for nitrate as the sole nitrogen source. The potential harm lies in the possible accumulation of the two toxic products of the pathway, nitrite and ammonium, when an alternate nitrogen source is available.

Thus the nitrate pathway is a normal point of contact between the environment and the metabolism of the plant. It is a prime port of entry for nitrogen, but not a unique one, since plants can utilize other nitrogen sources. Furthermore, nitrogen does not merely act as substrate which supports the growth of plants since the level of nitrogen available and the chemical form of the nitrogen can influence the developmental behavior of plants. These properties make the nitrate pathway an appropriate place for a plant to gather, analyze and respond to exogenous and endogenous signals concerning nitrogen supplies and requirements. They also make the nitrate pathway an attractive system in which to study the handling of environmental information.

The pathway is equally important in lower forms which can assimilate nitrate nitrogen. It has been studied in bacteria, fungi, and algae, as well as in flowering plants (Hewitt and Nicholas, 1964; Hattori and Uesugi, 1968; Beevers and Hageman, 1969). The similarities from organism to organism are sufficiently great to make this wealth of information useful, regardless of the organism under investigation.

Not the least of the attractions of this pathway in higher plants is that it is regulated. So far, it is the best example in higher plants of pathway activities that are regulated as a set (Filner *et al.*, 1969). Furthermore, the pathway is probably universal in higher plants (Sanderson and Cocking, 1964; Beevers *et al.*, 1964; Maretzki and Dela Cruz, 1967), and it is probably universally regulated, too (Afridi and Hewitt, 1964; Ferrari and Varner, 1969; Ingle *et al.*, 1966; Stewart, 1968).

Very little is known about the nitrate transport system even in bacteria and fungi. Nitrate reduction is by far the most completely characterized step. Compared to it, the information available about nitrite reduction is rudimentary.

NITRATE REDUCTASE AND NITRITE REDUCTASE ENZYMOLOGY

The current picture (Fig. 1) of the properties of nitrate reductase developed largely out of studies of the enzyme from *Neurospora crassa* (Nicholas and Nason, 1954). It was recently purified to crystallinity from this source (Garrett and Nason, 1969). The *Neurospora* enzyme contains molybdenum and a cytochrome of the *b* type. NADPH is the preferred reductant, and FAD is required as a cofactor. The enzyme also has FAD-dependent NADPH:cytochrome *c* reductase activity, which involves only the first part of the enzyme's electron transport chain, and methyl viologen:nitrate reductase activity, which involves only the last part of the electron transport chain. The first part is more sensitive than the last part to heat, SH poisons, and metal binding agents. The activity of the second part can be lost from fungi through mutation without loss of the first part (Sorger and Giles, 1965; Pateman *et al.*, 1964).

Under anaerobic conditions, *Escherichia coli* and other bacteria develop a nitrate reductase activity with an apparent respiratory rather than assimilatory function (Hewitt and Nicholas, 1964). In these bacteria, nitrate is believed to be used as a substitute terminal oxidant in place of oxygen. These nitrate reductases are particulate and usually have an associated cytochrome of the *b* type. The particles from *E. coli* (Taniguchi and Itagaki, 1960) will use NADH, for-

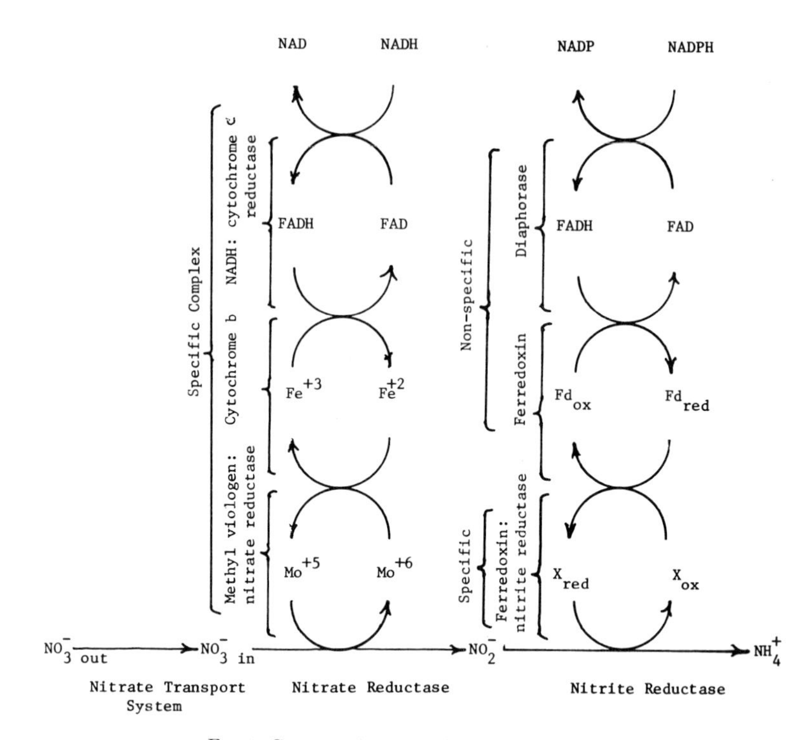

FIG. 1. Current picture of the nitrate pathway.

mate, FADH, or reduced methylviologen as reductant. A partially solubilized preparation that still has associated cytochrome b_1 will no longer use formate of NADH as reductant. Further solubilization and removal of the cytochrome results in isolation of a molybdenum containing protein which will use only reduced methylviologen as reductant. Thus the respiratory nitrate reductase of bacteria is very similar to the assimilatory nitrate reductase of fungi.

The nitrate reductase of higher plants has not been purified sufficiently from any source to show rigorously that the enzyme has bound FAD, molybdenum, and a cytochrome *b*. However, the enzyme from some sources shows some stimulation in response to added FAD (Evans and Nason, 1953; Maretzki and Dela Cruz, 1967; Spencer, 1959) and molybdenum (Maretzki and Dela Cruz, 1967). Also, radioactive molybdenum was reported to be associated with nitrate reductase activity through several purification steps (Evans and Hall, 1955), and it is required for induction of nitrate reductase in molybdenum-starved plants (Afridi and Hewitt, 1964).

Dr. John Wray of our laboratory has gathered data on the nitrate reductase of barley plants which illustrate the extent of the similarity of the plant enzyme to the bacterial and fungal enzymes (Wray *et al.*, 1969). Associated with barley NADH:nitrate reductase are NADH:cytochrome *c* reductase and FMNH:nitrate reductase activities. All three activities sediment in a common band in a sucrose gradient, with a sedimentation coefficient of very close to 8 S. The purified *Neurospora* enzyme has the same sedimentation coefficient (Garrett and Nason, 1969). The NADH:cytochrome *c* reductase is more heat labile and more sensitive to SH poisons than is the FMNH:nitrate reductase. Tungstate, which is an analog of molybdate, inhibits the development of FMNH:nitrate reductase, but not that of NADH:cytochrome *c* reductase, and molybdate antagonizes this inhibition (Heimer *et al.*, 1969).

The properties of the various nitrate reductases are consistent with the following scheme of electron transfer:

$$\text{NADH} \rightarrow \text{FAD} \rightarrow \text{Fe} \rightarrow \text{Mo} \rightarrow \text{NO}_3$$

The typical nitrate reductase probably consists of a complex of at least three protein moieties: cytochrome reductase, cytochrome, and nitrate reductase (Fig. 1). Recent genetic data on *E. coli* suggest that there are only these three kinds of components in the complex (Ruiz-Herrera *et al.*, 1969).

Nitrite reductases have been prepared from bacteria, fungi (Hewitt and Nicholas, 1964), algae (Hattori and Uesugi, 1968), and flowering plants (Joy and Hageman, 1966). The enzyme from green tissue of flowering plants is associated with chloroplasts (Ritenour *et al.*, 1967). However, nongreen tissues can also reduce nitrate. Ferredoxin is believed to be the physiological reductant (Joy and Hageman, 1966). Reduced methylviologen is also a suitable reductant. The prosthetic groups, electron transfer chain, and bound intermediates between nitrite and ammonium are not definitely known.

REGULATION OF THE NITRATE PATHWAY

Nitrate has been shown to induce the development of nitrate reductase in a wide variety of flowering plants (Beevers and Hageman, 1969). Nitrate also induces development of nitrite reductase. Inhibitors of protein synthesis inhibit the development of these enzymes, suggesting a dependence on protein synthesis, but not necessarily *de novo* synthesis of reductase proteins. The molecular origin of the induced activities remains to be determined.

In addition to nitrate, the development of nitrate reductase in leaves has been shown to depend on adequate levels of light, oxygen, CO_2, temperature, and molybdenum (Filner *et al.*, 1969). Nitrate cannot induce development of the enzyme in molybdenum-deficient plants (Afridi and Hewitt, 1964). This may simply reflect the fact that molybdenum is a component of nitrate reductase. However, the kinetics of enzyme development, when plants starved for molybdenum but not nitrate are exposed to molybdate, suggest that molybdenum may be a regulator of nitrate reductase formation as well as a component of the enzyme.

Nitrate reductase activity decays rapidly *in vivo* when plants are shifted from inducing to noninducing conditions. The level of nitrate reductase may therefore be a steady-state level determined by the relative rates of formation and decay of the enzyme, even under inducing conditions. It remains to be determined whether increases in enzyme activity are due to increases in rate of formation or decreases in rate of decay.

Most attempts to demonstrate inhibition of development of nitrate reductase by ammonium or amino acids in flowering plants have been unsuccessful (Beevers and Hageman, 1969). However, ammonium inhibits development of nitrate reductase in algae and fungi. Under conditions similar to those used to culture algae and fungi, amino acids inhibit development of nitrate reductase in tobacco cells cultured *in vitro* (Filner, 1966).

Mutants of *E. coli* (Venables *et al.*, 1968), *Aspergillus* (Cove and Pateman, 1969), and *Neurospora* (Sorger and Giles, 1965) have the odd characteristic that the NAD(P)H:cytochrome *c* reductase portion of nitrate reductase, and in *Aspergillus*, nitrite reductase as well, are constitutive if functional nitrate reductase has been lost. This has been interpreted to mean that functional nitrate reductase plays a direct role in repression of the nitrate pathway enzymes (Cove and Pateman, 1969). Apparently, the mechanism governing the development of the nitrate pathway in these organisms has some complications not encountered in the more extensively studied pathway regulation systems of bacteria.

THE CULTURED TOBACCO CELL SYSTEM

During the course of a study of the control of nitrate reductase in cultured tobacco cells, indications were found that the regulatory mechanism could function in two modes: as either a conventional negative feedback loop, or as an on-off switch governing the growth of the cells (Filner, 1966). We have now more thoroughly examined the control of the pathway. It is this more recent work that I wish to discuss. Much of the progress made has been due to the efforts of Mr. Yair Heimer (1970), who has concentrated on the role of nitrate uptake in the regulation of the nitrate pathway, both in tobacco XD cells, our wild type, and in an interesting variant which he isolated and characterized, the XDR^{thr} strain. Dr. Hanna Chroboczek-Kelker (1969) completed our coverage of the pathway by investigating the control of nitrite reductase. She has found nitrite reductase to be better suited in certain respects for regulation studies than nitrate reductase.

The cells are derived from pith cells of *Nicotiana tabacum* L. cv. Xanthi. They are grown in shake cultures on a medium containing inorganic salts, B vitamins, sucrose, and 2,4-dichlorophenoxyacetic acid. The cells multiply exponentially with a generation time of 2 days (Filner, 1965). Exponential growth ceases when the nitrogen available in the standard medium, 2.5 m*M* nitrate, is exhausted. A maximum yield of ca. 10^5 cells/ml or 40 mg fresh weight per milliliter is obtained. Nitrate can be replaced by urea, γ-aminobutyric acid or acid-hydrolyzed casein with little change in the generation time. Ironically, ammonium neither replaces nitrate nor affects growth on nitrate, a result suggesting that it may not penetrate into the cells under our culture conditions.

The cells grow in the form of filaments, with cell division occurring throughout the filament (Fig. 2a). In the exponential phase, mitosis seems to occur almost as soon as the longitudinal growth provides room to segregate two nuclei on either side of a cross wall. However, as the culture approaches the stationary phase, the volume of the average cell greatly increases, both by lateral and longitudinal expansion. The long filaments dissociate into filaments made up of fewer, but greatly expanded, cells (Fig. 2b).

In the earlier work on control of nitrate reductase, it was found that nitrate induces the development of active nitrate reductase (Fig. 3b), and that the enzyme decays *in vivo* (Fig. 4) as the medium is depleted of nitrate (Filner, 1966). Furthermore, casein hydrolyzate inhibits the development of nitrate reductase in proportion to its ability to replace nitrate as the nitrogen source. Thus in the presence of a balanced amino acid mixture, the regulatory mechanism functions as a classical negative feedback control device, its function being to maintain the optimal flux of reduced nitrogen.

Using reduced methylviologen as reductant and the disappearance

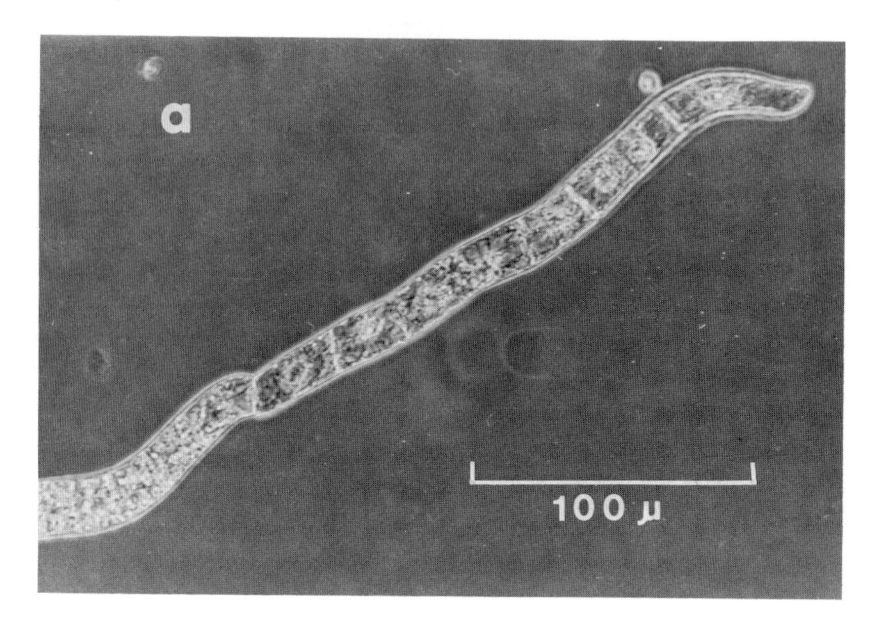

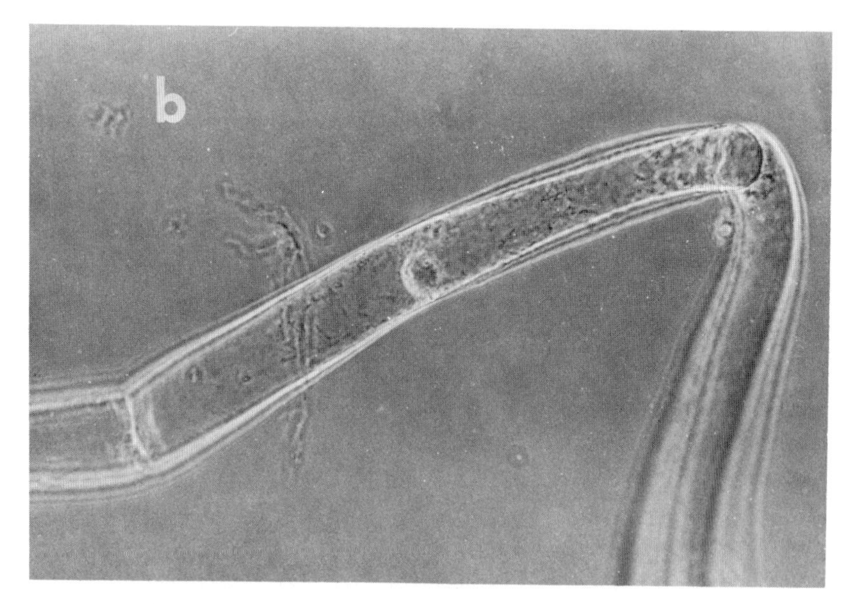

FIG. 2. Tobacco XD cells (a) from a culture in midexponential phase; (b) from a culture in stationary phase. The cells expand greatly late in the exponential phase and during the stationary phase, when grown on a medium in which nitrogen is limiting.

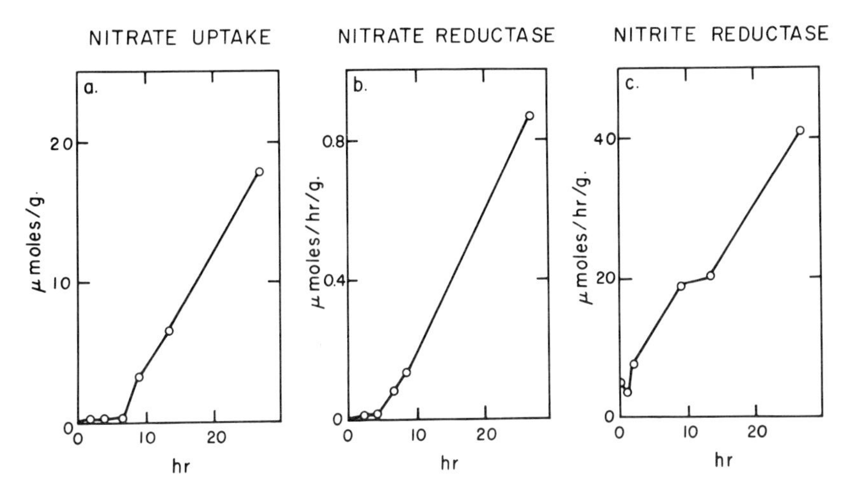

FIG. 3. Increases in nitrate pathway activities in response to nitrate. Tobacco XD cells in stationary phase on a medium they depleted of nitrate were subcultured into new medium rich in nitrate. At various times, cells were harvested and homogenized. Nitrate (a), nitrate reductase (b), and nitrite reductase (c) were assayed. All three parameters rise markedly, but each with a different time dependence. From Heimer (1970) and Chroboczek-Kelker (1969).

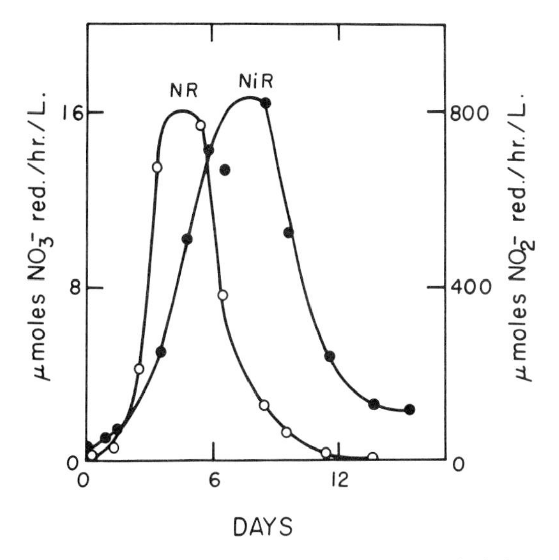

FIG. 4. Variation with culture age of nitrate reductase and nitrite reductase in extracts of tobacco XD cells. Initial conditions as in Fig. 3. The cells consume the nitrate by the tenth day. Both activities rise, then decay, but nitrite reductase begins to decay later. From Chroboczek-Kelker (1969).

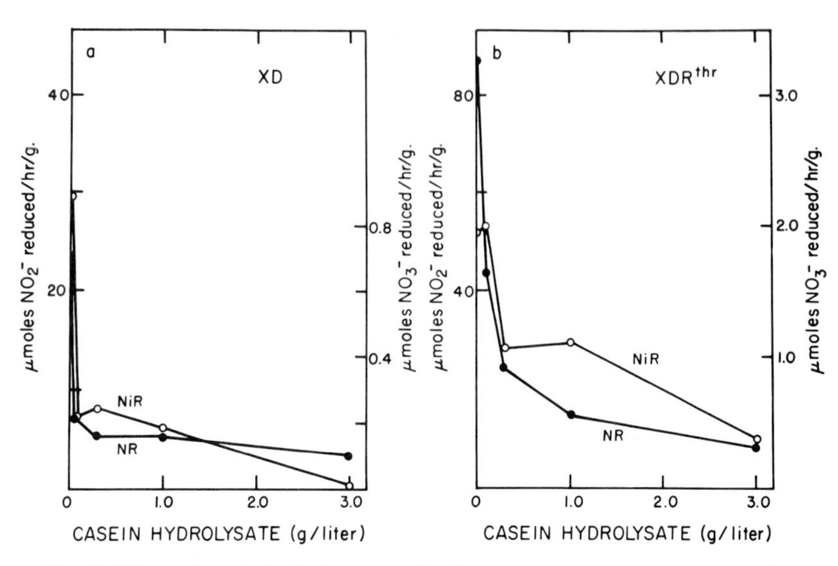

FIG. 5. Effect of casein hydrolyzate on development of nitrate reductase in (a) XD and (b) XDR^{thr} cells. Initial conditions as in Fig. 3, except that the medium was supplemented with acid-hydrolyzed casein where indicated. The enzyme activities were determined after 48 hours. The development of both enzymes is inhibited in both cell lines, with similar dependences on concentration of casein hydrolyzate. The activity levels of both enzymes are higher in the XDR^{thr} cells. From Chroboczek-Kelker (1969) and Heimer (1970).

of nitrite as an assay, it has been determined that the development of active nitrite reductase is induced by nitrate (Fig. 3c); nitrite reductase decays as the medium is depleted of nitrate, but somewhat later than nitrate reductase (Fig. 4); and the development of nitrite reductase is inhibited by casein hydrolyzate (Fig. 5a) (Chroboczek-Kelker, 1969). Nitrate reductase and nitrite reductase thus behave quite similarly, but they are not two activities of the same enzyme. They can be physically separated, for instance, by ammonium sulfate fractionation.

We have come to the conclusion that the development of the nitrate transport system is also controlled by nitrate and amino acids. Nitrate uptake was studied by assaying the nitrate accumulated in previously nitrate-free cells which were obtained by allowing cells to grow into the stationary phase under nitrate-limited conditions. It was first established that both efflux and reduction of nitrate are negligible with respect to the amount of nitrate accumulated during the first 20 hours or so. Under these circumstances the rate of nitrate accumulation is a useful measure of the initial rate of uptake. The kinetics of nitrate accumulation consist of two phases (Fig. 3a). For the first 6 hours, nitrate enters slowly, and the internal concentration does not exceed the external concentration. Then a phase of constant, rapid nitrate uptake abruptly begins. After about 24 hours, the steady-state level of accumulated nitrate is reached. It exceeds the external level by 10–20 times. The lag does not represent the time required to overcome the effects of nitrogen starvation. Essentially the same kinetics of nitrate accumulation are obtained with exponential cells on urea which are shifted to nitrate medium. Such cells have not been subjected to nitrogen starvation. Therefore, the development of the capacity for rapid nitrate uptake seems to be induced by nitrate. However, the kinetics do not quite fit such a simple picture. Rather, they suggest that after a lag of 6 hours, the uptake system is very rapidly activated, since the maximum uptake rate develops abruptly. Nevertheless, nitrate starts the process, whatever it is. As in the cases of nitrate reductase and nitrite reductase, casein hydrolyzate inhibits the development of the rapid uptake system (Fig. 6).

CONTROL OF GROWTH THROUGH THE NITRATE PATHWAY

Now let us return to the observations which suggested that growth of tobacco cells could be inhibited through the functioning of the regulatory mechanism which governs the nitrate pathway (Filner, 1966). At concentrations well below substrate levels, some individual amino acids inhibited the development of nitrate reductase, while a few, notably arginine, lysine, cysteine, and isoleucine, did not inhibit. The noninhibitors antagonized the inhibitors. Furthermore, the amino acids which inhibited development of nitrate reductase also inhibited growth, and the antagonistic amino acids also antagonized the inhibition of growth.

Based on these observations, it was hypothesized that when the tobacco XD cell encounters single amino acids, its control mechanism for the nitrate pathway behaves in a manner which is illogical for a classical negative feedback device: in spite of the inadequacy of the single amino acid at low concentration as a nitrogen source, the control mechanism seems to shut down the development of the nitrate pathway activities. This in turn lowers the flux of reduced nitrogen through the pathway, so that growth is inhibited due to starvation

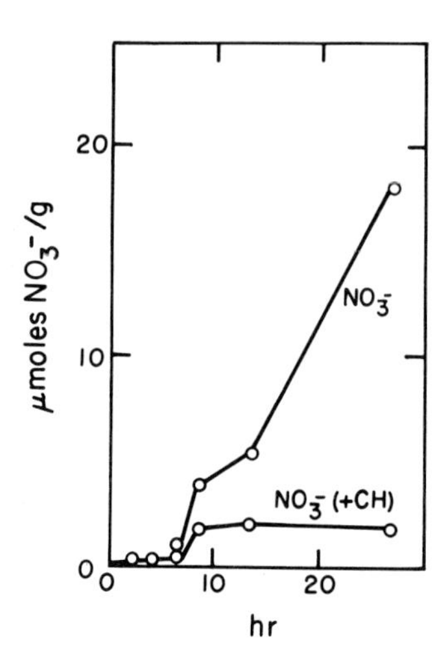

FIG. 6. Effect of casein hydrolyzate on accumulation nitrate by nitrogen starved cells. Initial conditions as in Fig. 5. It inhibits accumulation of nitrate. From Heimer (1970).

for reduced nitrogen. This hypothesis has now been substantiated by two lines of evidence (Heimer and Filner, 1969).

First, cells grown on urea or γ-aminobutyric acid are not inhibited by concentrations of single amino acids which inhibit growth on nitrate. Therefore, the mechanism of growth inhibition works only when the cells are dependent upon the nitrate pathway for growth.

The second approach has been through the isolation and characterization of a variant cell line, XDR^{thr}, which was selected for resistance to threonine. Threonine is among the most potent of the growth-inhibiting amino acids. It was reasoned that if the hypothesis were correct, it might be possible by mutation to alter the sensitivity of the regulatory system to the inhibitory amino acids. Variant cells with this property could be selected by their ability to grow in the presence of a normally inhibitory level of an amino acid. A change in the regulatory mechanism is not the only way that a sensitive cell might become resistant, but as I think you will see, the properties of the XDR^{thr} cells are consistent with a regulatory change.

The variant was obtained from the XD line by exposure of cells to nitrosoguanidine, followed by growth in a sublethal level of threonine, and then finally repeated selection for growth in the presence of a lethal level of threonine (Heimer and Filner, 1969). Subsequent experiments indicated that a more direct selection could be made by exposing XD cells directly to a lethal level of threonine. The resistant character of the XDR^{thr} cells was inherited through more than 40 doublings in the absence of the selective agent. Therefore, resistance is not due to a reversible adaptation, but rather to mutation, or possibly a metastable adaptation (differentiation?).

When grown on the standard medium, the XDR^{thr} cells have a level of extractable nitrate reductase which is higher than that in XD cells (Table I and Fig. 5b). Nitrite reductase is also elevated in XDR^{thr} cells. This alone would be a very improbable finding if the inhibition of growth by amino acids were not linked to their role as effectors in the nitrate pathway regulatory system.

The XDR^{thr} cells are resistant to several of the inhibitory amino acids besides threonine. This suggests that resistance is due to a change in a common mechanism by which the various amino acids inhibit. This interpretation is supported by the fact that threonine-^{14}C enters and accumulates as threonine in both XD and XDR^{thr} cells, so that resistance is not due to exclusion or degradation of the inhibitor (Heimer and Filner, 1969).

We were surprised to find that, although nitrate reductase and nitrite reductase are elevated in XDR^{thr} cells, the development of both enzymes is as sensitive to casein hydrolyzate as in the XD cells (Fig. 5). Clearly, the feedback control mechanism is still quite functional in the XDR^{thr} cells as far as the two reductases are concerned.

TABLE 1

COMPARISON OF NITRATE PATHWAY CONTROLS IN XD AND XDR^{thr} CELLS[a, b]

Medium nitrogen	Nitrate accumulation (μmoles/gm/24 hr)		Nitrate reductase (units[c]/gm/24 hr)	
	XD	XDR^{thr}	XD	XDR^{thr}
2.5 m*M* nitrate	46.2	54.5	0.715	3.30
2.5 m*M* nitrate + 0.1 gm/liter casein hydrolyzate	16.4	43.8	0.215	1.64
2.5 m*M* nitrate + 0.3 gm/liter casein hydrolyzate	5.7	28.5	0.155	0.900

[a] Heimer (1970).

[b] Initial conditions as in Fig. 5.

[c] One unit of nitrate reductase reduces 1 μmole of nitrate per hour under the assay conditions.

The picture is quite different for the nitrate uptake system. Casein hydrolyzate still inhibits nitrate uptake in XDR^{thr} cells, but far less effectively than in the XD cells (Table I). Thus we found that resistance to the growth-inhibiting amino acids was associated in the XDR^{thr} cells with resistance to inhibition of nitrate uptake by amino acids. Because of the improbability of obtaining a double variant in two independent properties, the characteristics of the XDR^{thr} cells strongly support the hypothesis that amino acids inhibit growth by inhibiting the development of the nitrate pathway.

CONTROL CIRCUITRY OF THE NITRATE PATHWAY

The regulatory relationships of the nitrate pathway in tobacco XD cells are summarized in Fig. 7. The nitrate pathway activities—nitrate uptake, nitrate reductase, and nitrite reductase—are absent in cells grown on media lacking nitrate. If nitrate is present, all three activities can develop.

Not all the nitrate which accumulates is either necessary to induce the development of nitrate reductase or sufficient to maintain it. This becomes apparent from the kinetics of development of nitrate reductase in exponential cells taken from urea medium and shifted to nitrate medium (Fig. 8). The enzyme begins to develop immediately, before any accumulation of nitrate has occurred. Thus very little nitrate is required to induce. Furthermore, if fully induced cells loaded with nitrate at 10 times or more the external concentration are shifted to nitrateless medium, the enzyme begins to decay after

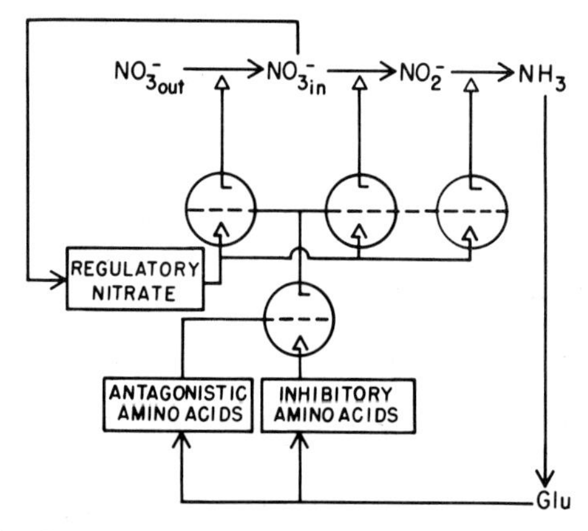

FIG. 7. Functional and control relationships of the nitrate pathway in tobacco cells.

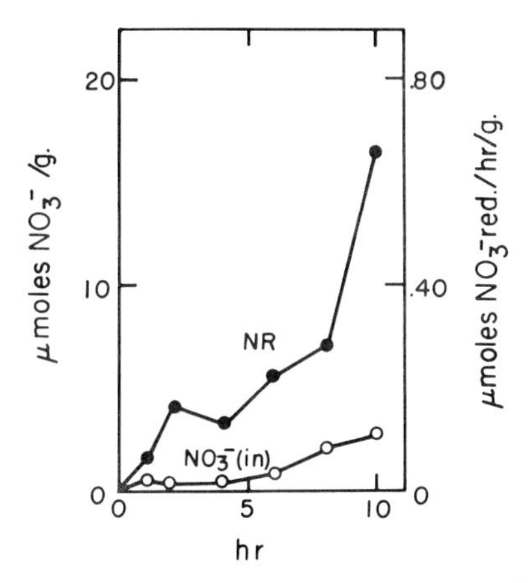

FIG. 8. Accumulation of nitrate and development of nitrate reductase in tobacco XD cells not previously starved for nitrogen. The cells were grown on urea as sole nitrogen source, then were transferred in the exponential phase to medium with nitrate, but no urea. Nitrate reductase begins to develop immediately. Very little nitrate need be in the cells to induce development of nitrate reductase. From Heimer (1970).

only a few percent of the internal nitrate has been consumed. These observations lead to the conclusion that only a small portion of the nitrate in the cells induces, and the portion which induces, or maintains the induced state, has a very short half-life compared to either the bulk of the endogenous nitrate or even the nitrate reductase.

The effectiveness of nitrate as inducer is modulated by amino acids. One group of amino acids inhibits the development of nitrate reductase. This effect can be schematically represented as a grid potential which attenuates the inducing potential of nitrate. The effect of a second group of amino acids which antagonizes the effect of the first group can also be represented as an attenuating grid potential.

The XDR^{thr} cells have a normal control system for nitrate reductase and nitrite reductase, but the control for nitrate uptake is reduced in sensitivity to inhibitory amino acids. Therefore, the attenuation of induction of nitrate uptake is diagrammed as being separable from the attenuation of induction of nitrate reductase and nitrite reductase.

The controls of nitrate uptake, nitrate reductase, and nitrite reductase development are depicted as being free of any direct dependence on the functioning of steps in the pathway. This is in contrast to the picture derived from genetic studies of *Aspergillus*, in which there appears to be a dependence on functional nitrate reductase (Cove and Pateman, 1969). The primary basis for believing that functional nitrate reductase is not an integral part of the control in tobacco cells is that inhibition of the development of functional nitrate reductase with tungstate does not interfere with the development of either nitrate uptake or nitrite reductase, nor the effectiveness of casein hydrolyzate as inhibitor of development of nitrate uptake or nitrite reductase (Chroboczek-Kelker, 1969; Heimer *et al.*, 1969; Heimer, 1970).

The diagram of the regulatory relationships is consistent with the bulk of the observations. However, the actual mechanism for increasing or decreasing the activity levels is not known. It still remains to be determined what the relative contributions are from enzyme activation, inhibition, synthesis, and degradation. One of the most striking properties of the control system governing nitrate reductase is the decay of the enzyme activity following removal of nitrate or addition of casein hydrolyzate (Fig. 9). We do not know whether the enzyme is always decaying at the high rate observed under these conditions, or if the decay rate is greatly accelerated by the conditions. Nevertheless, the decay, whether it is a reflection of a modifi-

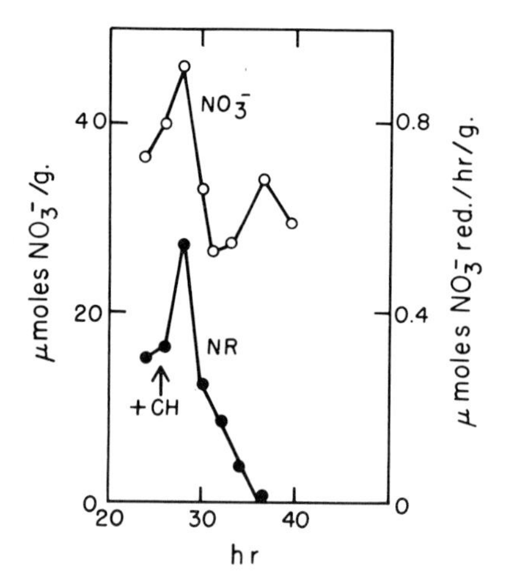

FIG. 9. Effect of casein hydrolyzate on nitrate reductase after it has formed. Initial conditions as in Fig. 3. At the indicated time, casein hydrolyzate was added to the medium. The enzyme activity extractable from the cells began to decrease shortly thereafter. The high accumulation of nitrate within the cells did not prevent the decay. From Heimer (1970).

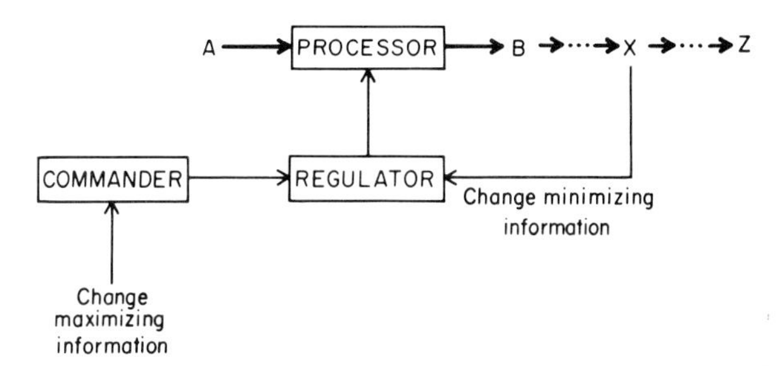

FIG. 10. A negative feedback loop is normally a mechanism for resisting change in a parameter (rate of production of z in the diagram). It can become a mechanism for promoting change if there is a way to alter the command value for the regulator. The existence of a commander may be inferred from anomalous behavior of a negative feedback loop. Such a mechanism seems to exist in the control circuitry of the nitrate pathway in tobacco cells.

cation of the catalytic site or of complete proteolysis of the enzyme, is clearly a major factor in determining the *in vivo* level of nitrate reductase.

The available evidence strongly supports the contention that the control system governing the nitrate pathway can operate in two modes. In the first mode, it is a classical negative feedback device which maintains the rate of supply of reduced nitrogen at that which is required to support the maximum possible proliferation rate of the tobacco cells. In the second mode, it appears to determine the proliferation rate by greatly lowering the rate of supply of reduced nitrogen. This latter mode of operation may be an example of Nature run amok in the culture flask. On the other hand, it may be a manifestation of a physiologically important property of the control system.

CONCLUDING REMARKS

Many cells of multicellular organisms do not proliferate at the maximum rate possible in their environment. When they do, the phenomenon is called cancer. Therefore, there must exist mechanisms which enable the cells of multicellular organisms to disregard the richness of their environment, so that they can function for the greater good of the organism as a whole. Such a mechanism would not be expected in unicellular organisms because what is best for propagation of the cell is synonymous with what is best for propagation of the organism. A bacterial cell, for instance, would be expected always to respond to its environment in a manner which tends to promote the maximum proliferation rate possible. Apparent exceptions to this expectation are known, however (Burlant *et al.*, 1965). Perhaps the fundamental innovation of cells with the developmental capability is exploitation of the mechanisms which enable them to escape from the unicellular behavior pattern of multiplying as fast as possible in a given environment. Let us call this mechanism the commander, because it has the function of providing a regulatory device with a command value other than that which is most compatible with the maximum rate of proliferation (Fig. 10). If this idea is approximately correct, then in the control of the nitrate pathway of cultured tobacco cells we have perhaps had a glimpse of a commander in operation. We are mildly hopeful that the nitrate pathway mechanism will prove to be a fruitful model to study. Others will probably be found, or possibly are already known.

ACKNOWLEDGMENT

The hard experimental facts in this paper are drawn from the individual research efforts of my colleagues, Yair M. Heimer, Hanna Chroboczek-Kelker, and John L. Wray. The ideas and interpretations presented in this synthesis are my summary of the countless discussions we have had. This research was supported by U.S. Atomic Engergy Commission Contract AT-(11-1)-1338.

REFERENCES

Afridi, M. M. R. K., and Hewitt, E. J. (1964). The inducible formation and stability of nitrate reductase in higher plants. I. Effects of nitrate and molybdenum on enzyme activity in cauliflower (*Brassica oleracea* var. *botrytis*) *J. Exptl. Bot.* **15,** 251–271.

Allsopp, A. (1964). The metabolic status and morphogenesis. *Phytomorphology* **14,** 1–10.

Bandurski, R. S. (1965). Biological reduction of sulfate and nitrate. *In* "Plant Biochemistry" (J. Bonner and J. E. Varner, eds.), 2nd ed., pp. 467–490. Academic Press, New York.

Beevers, L., and Hageman, R. H. (1969). Nitrate reduction in higher plants. *Ann. Rev. Plant Physiol.* **20,** 495–522.

Beevers, L., Flesher, D., and Hageman, R. H. (1964). Studies on the pyridine nucleotide specificity of nitrate reductase in higher plants and its relationship to sulfhydryl level. *Biochim. Biophys. Acta* **89,** 453–464.

Burlant, L., Datta, P., and Gest, H. (1965). Control of enzyme activities in growing bacterial cells by concerted feedback inhibition. *Science* **148,** 1351–1353.

Chroboczek-Kelker, H. (1969). Regulation of nitrite reductase in tobacco cells. Ph.D. Thesis, Part 2, Michigan State University, East Lansing.

Cove, D. J., and Pateman, J. A. (1969). Autoregulation of the synthesis of nitrate reductase in *Aspergillus nidulans*. *J. Bacteriol.* **97,** 1374–1378.

Evans, H. J., and Hall, N. E. (1955). Association of molybdenum with nitrate reductase from soybean leaves. *Science* **122,** 922–923.

EVANS, H. J., and NASON, A. (1953). Pyridine nucleotide-nitrate reductase from extracts of higher plants. *Plant Physiol.* **28,** 233–254.

FERRARI, T. E., and VARNER, J. E. (1969). Substrate induction of nitrate reductase in aleurone layers. *Plant Physiol.* **44,** 85–88.

FILNER, P. (1965). Semiconservative replication of DNA in a higher plant cell. *Exptl. Cell Res.* **39,** 33–39.

FILNER, P. (1966). Regulation of nitrate reductase in cultured tobacco cells. *Biochim. Biophys. Acta* **118,** 299–310.

FILNER, P., WRAY, J. L., and VARNER, J. E. (1969). Enzyme induction in higher plants. *Science* **165,** 358–367.

FRIES, N. (1954). Chemical factors controlling the growth of decotylized pea seedlings. *Symbolae Botan. Upsalienses* **13,** 1–83.

GARRETT, R. H., and NASON, A. (1969). Further purification and properties of *Neurospora* nitrate reductase. *J. Biol. Chem.* **244,** 2870–2882.

HATTORI, A., and UESUGI, I. (1968). Purification and properties of nitrate reductase from the blue-green alga, *Anabaena cylindrica*. *Plant Cell Physiol.* **9,** 689–699.

HEIMER, Y. M. (1970). Ph.D. thesis, Michigan State University, East Lansing, in preparation.

HEIMER, Y. M., and FILNER, P. (1969). Regulation of nitrate reduction in cultured tobacco cells. II. Isolation and partial characterization of a variant cell line. In preparation.

HEIMER, Y. M., WRAY, J. L., and FILNER, P. (1969). The effect of tungstate on nitrate assimilation in higher plant tissues. *Plant Physiol.* **44,** 1197–1199.

HEWITT, E. J., and NICHOLAS, D. J. D. (1964). Enzymes of inorganic nitrogen metabolism. *In* "Modern Methods of Plant Analysis" (H. F. Linskens, B. D. Sanwal, and M. V. Tracey, eds.), Vol. 7, pp. 67–172. Springer, Berlin.

INGLE, J., JOY, K. W., and HAGEMAN, R. H. (1966). The regulation of activity of the enzymes involved in the assimilation of nitrate by higher plants. *Biochem. J.* **100,** 577–588.

JACOB, F., and MONOD, J. (1963). Genetic repression, allosteric inhibition, and cellular differentiation. *In* "Cytodifferentiation and Macromelecular Synthesis" (M. Locke, ed.), pp. 30–64. Academic Press, New York.

JOY, K. W., and HAGEMAN, R. H. (1966). The purification and properties of nitrite reductase from higher plants, and its dependence on ferredoxin. *Biochem. J.* **100,** 263–273.

KESSLER, E. (1964). Nitrate assimilation by plants. *Ann. Rev. Plant Physiol.* **15,** 57–71.

MARETZKI, A., and DELA CRUZ, A. (1967). Nitrate reductase in sugarcane tissues. *Plant Cell Physiol.* **8,** 605–611.

NICHOLAS, D. J. D., and NASON, A. (1954). Mechanism of action of nitrate reductase from *Neurospora*. *J. Biol. Chem.* **211,** 183–197.

PATEMAN, J. A., COVE, D. J., REVER, B. M., and ROBERTS, D. B. (1964). A common cofactor for nitrate reductase and xanthine dehydrogenase which also regulates the synthesis of nitrate reductase. *Nature* **201,** 58–60.

RITENOUR, G. L., JOY, K. W., BUNNING, J., and HAGEMAN, R. H. (1967). Intracellular localization of nitrate reductase, nitrite reductase and glutamic dehydrogenase in green leaf tissue. *Plant Physiol.* **42,** 233–237.

RUIZ-HERRERA, L., SHOWE, M. K., and DEMOSS, J. A. (1969). Nitrate reductase complex of *Escherichia coli* K12: isolation and characterization of mutants unable to reduce nitrate. *J. Bacteriol.* **97,** 1291–1297.

SANDERSON, G. W., and COCKING, E. C. (1964). The enzymic assimilation of nitrate in the tomato plant. I. Nitrate reductase. *Plant Physiol.* **39,** 416–422.

SINNOTT, E. W. (1960). "Plant Morphogenesis," pp. 363–373. McGraw-Hill, New York.

SORGER, G. J., and GILES, N. H. (1965). Genetic control of nitrate reductase in *Neurospora crassa*. *Genetics* **52,** 777–788.

SPENCER, D. (1959). A DPNH-specific nitrate reductase from germinating wheat. *Australian J. Biol. Sci.* **12,** 181–196.

STEINBERG, R. A. (1949). Symptoms of amino acid action on tobacco seedlings in aseptic culture. *J. Agr. Res.* **78,** 733–741.

STEWART, G. R. (1968). The effect of cycloheximide on the induction of nitrate and nitrite reductase in *Lemna minor* L. *Phytochemistry* **7,** 1139–1142.

TANIGUCHI, S., and ITAGAKI, E. (1960). Nitrate reductase of nitrate respiration type from *Escherichia coli*. I. Solubilization and purification from the particulate system with molecular characterization as a metalloprotein. *Biochim. Biophys. Acta* **44,** 263–279.

VENABLES, W. A., WIMPENNY, J. W. T., and COLE, J. A. (1968). Enzymic properties of of a mutant of *Escherichia coli* K12 lacking nitrate reductase. *Arch. Mikrobiol.* **63,** 117–121.

WALLACE, W., and PATE, J. S. (1965). Nitrate reductase in the field pea. *Ann. Botany* **29,** 655–671.

WALLACE, W., and PATE, J. S. (1967). Nitrate assimilation in higher plants with special reference to the cocklebur (*Xanthium pennsylvanicum* Wallr.) *Ann. Botany* **31,** 213–228.

WARIS, H. (1959). Neomorphosis in seed plants induced by amino acids. I. *Oenanthe aquatica*. *Physiol. Plantarum* **12,** 753–766.

WARIS, H. (1962). Neomorphosis in seed plants induced by amino acids. II. *Oenanthe lachenalii*. *Physiol. Plantarum* **15,** 736–753.

WRAY, J. L., FILNER, P., and RIES, S. K. (1969). The effect of tungstate on nitrate reductase in barley shoots. *Abstr. 11th Intern. Botani. Congr., Seattle, Wash., Aug./Sept. 1969*, p. 243.

DEVELOPMENTAL BIOLOGY SUPPLEMENT 3, 227–243 (1969)

Light in Plant and Animal Development

STERLING B. HENDRICKS

Soil and Water Conservation Research Division, Agricultural Research Service, USDA, Beltsville, Maryland

INTRODUCTION

Light and darkness are major environmental factors affecting plant and animal development. Day-length-dependent, or photoperiodic, controls of reproduction are main aspects of the subject. They include control of flowering in plants and reproduction in animals; in both these phenomena endogenous rhythmic changes are involved in addition to the response to light. Several responses dependent on photosensitized oxidations, and light action in the control of diapause of insects are also considered. Vision and photosynthesis, the two main functions involving light in life, are not discussed.

The fitness of visible light for coupling adaptation of a plant or animal to an environment rests in part on the pervasiveness of light and of the diurnal cycle of day and night since the first appearance of living things. Light can enter an object without disturbing other than specific, absorbing molecules, possibly deep in the target tissue. Low-energy radiations in the visible and the bordering, (near) ultraviolet and infrared regions cause only single excitations of the absorbing pigment molecules. The effective excitations are of electrons from lower to higher energy levels, rather than ionization as induced by high-energy radiation (X-rays, etc.) or only rotations and vibrations of constituent groups resulting from absorption of far-infrared radiation (heat). An electronic excitation is a unique molecular event at a specific point in a group of molecular determinants of a display. It is, in fact, likely to be the initiator of the chain of events leading to display, as is the case for vision.

Darkness is the absence of light of adequate energy for electronic excitation of reginal pigments. The most important property of darkness for the organism is duration during which thermal or dark reactions might proceed that otherwise are inhibited by light.

CONTROL OF PLANT REPRODUCTION BY LIGHT

I first treat photoperiodic responses of flowering in which light received by a leaf controls the differentiation of a terminal shoot meristem. The control is illustrated by development of the terminal spike or flowering structure of barley, which belongs to the general class of plants requiring long days to flower. When days are short the terminal shoot meristem of barley develops as an elongated, conelike structure bearing shelflike leaf primordia with no evidence of flowers (Fig. 1). If barley plants growing on short days are transferred to long ones or if long nights are interrupted by light, fuller spike formation is induced (Fig. 1). The degree of flowering display is used to assay the response to wavelength and energy of the light in the hope of finding something about the initiating reaction. The action spectrum (Fig. 2) shows a maximum effectiveness in the red part of the spectrum, at 600–680 nm. The response at any given wavelength depends on the radiant energy. It is half saturated at very low energy levels, namely 1.1×10^{-8} Einstein/cm^2 of incident energy, at 650 nm. Less than 1% of the incident light was later found to be absorbed in the effective system. Accordingly, about 10^{-10} mole of light-induced substance in the leaf controls flower differentiation at the terminal meristem.

FIG. 1. Development of the flowering spike of barley (*Hordeum vulgare* cv. Wintex) with increase in numbers of 16-hour days from none (left) to 21 (right). (Borthwick *et al.*, 1948.)

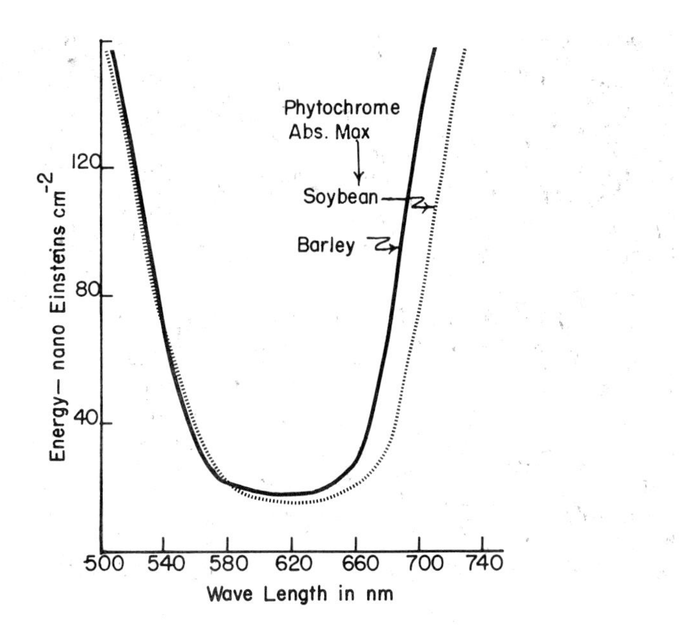

FIG. 2. Action spectra obtained with interruption by light of long nights for flower induction of barley and prevention of flowering of soybean. (Borthwick *et al.*, 1948.)

A plant of the general class requiring short days to flower remains vegetative when growing on long days or when long nights are interrupted by adequate light. The change in the degree of flowering is illustrated in Fig. 3 for *Chenopodium rubrum* (Kasperbauer *et al.*, 1963). Soybeans and chrysanthemums are also of this type. The action spectra for the two opposite types of plants, the long- and short-day ones, are closely the same as illustrated in Fig. 2 for soybean and barley. The responses of both types of plants to red light are reversed by far-red light with action maxima near 730 nm (Borthwick *et al.*, 1952). Energies required for half saturation in both directions of change are about the same. It follows that a reversible change in the form of a single effective pigment is involved. This can be written as,

$$P_r \underset{730\text{ nm}}{\overset{660\text{ nm}}{\rightleftharpoons}} P_{fr}$$

This reversibility of a potentiated differentiation, which is unique in the biological world, has many implications. An immediate one is

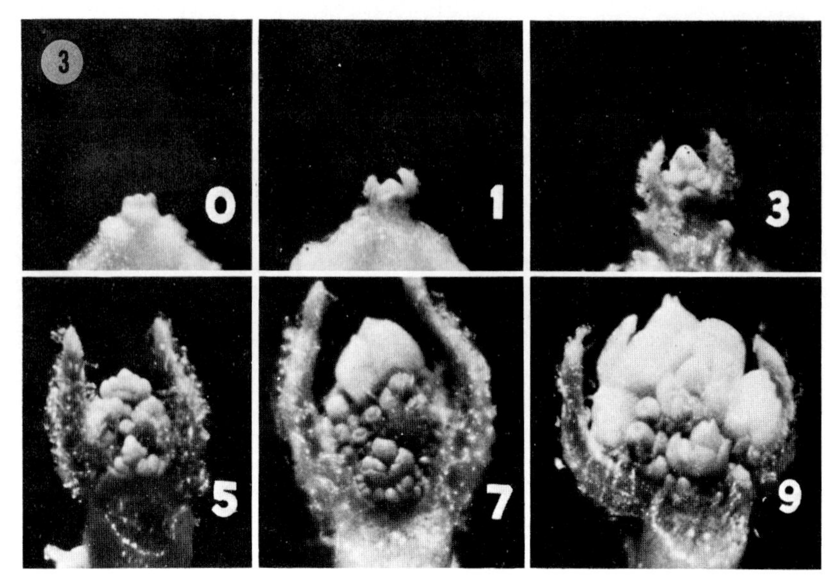

FIG. 3. Development of the terminal inflorescence of *Chenopodium rubrum* induced by decreasing irradiances in red light interruptions of long nights. Stage 9 is the dark control. (Kasperbauer *et al.*, 1963.)

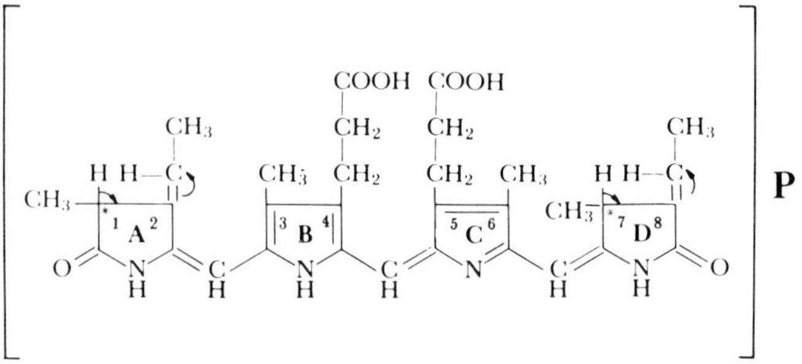

COOH COOH
CH₃ CH₂ CH₂ CH₃
H H—C CH₃ CH₂ CH₂ CH₃ H H—C
CH₃ A B C CH₃ D P
O N C N C N C N O
H H H H H H

FIG. 4. A probable formula for the bilitriene chromophore of phytochrome. Brackets and P indicate the associated protein. (H. W. Siegelman, personal communication.)

that it can be used for assay in attempts to isolate *P*. The attempts were successful and *P*, now known as phytochrome, was shown to be a photoreversible blue chromoprotein with a bilitriene (phycobilin) as the chromophore or active group (Fig. 4).

The environmental control of reproduction in plant phyla shows many nuances. In some lower plants, transition from asexual to sexual reproduction is under such control (Mohr, 1965). Among seed plants, the length of the day can control sexuality in the production of perfect flowers, which may be chasmogamous or cleistogamous, or unisexual ones as shown by hemp and Japanese hop. Apomixis, a type of development that does not involve meiosis, is also controlled, as in *Poa* species and many other grasses. Through apomixis, races of a species can remain genetically isolated in a favorable environment, but still be able to revert to sexual reproduction in changed surroundings. Various degrees of control exist over reproductive and vegetative development, which are responsive to changes in day length and variations in temperature.

The control of flowering in plants, which is a change in differentiation of a meristem, by change in day length involves a timing process, as implied by the term "photoperiodism." While light action is a part of the process it is not the only part. The timing also involves rhythmic changes as implied by the terms "biological clock" or "physiological clock" (Bünning, 1963). Rhythmic changes in plants can be seen in leaf movements, stomatal opening, and other responses as well as in flowering. Underlying causes for the rhythms, in a biochemical or physiological sense, are not known for plants, but are widely supposed to involve interplay of several hormonal actions. The light action couples with the rhythmic changes either by permitting their display or by shifting the onset of a rhythm—the so-called "phasing." Thus, one is left with some understanding of the initial photochemical events in plants and the elaboration of the actual flower development, but knows woefully little of what goes on in between.

OTHER RESPONSES OF PLANTS TO CHANGE IN FORM OF PHYTOCHROME

How does phytochrome act in controlling flowering? The question is raised both for the molecular level of action and in terms of the differentiation display. Pursuit of an answer leads in unanticipated directions before coming back toward photoperidism.

Plants show a number of responses to light other than control of flowering. These include partial inhibition of stem lengthening and control of germination of some seed, neither of which is periodic. The effective pigment system in the several cases turns out to be phytochrome, as shown by details of the action spectra and by red, far-red photoreversibility (Hendricks and Borthwick, 1965). The seed response is release of a dormancy, i.e., of a state of suspended growth. A very small change of P_r to P_{fr}, when P_r is very predominant, is strikingly more effective on germination and stem elongation than is the reverse change when P_{fr} is predominant. It follows that P_{fr}, the 730 nm absorption maximum form, is the biologically active one. Phytochrome, moreover, is active in a number of displays that are seemingly unrelated (Table 1). Many of these displays, such as axis modification and plastid orientation, do not involve differentiation.

Phytochrome in the P_{fr} form is also required to initiate development of the lamellar structure of plastids in tissue taken from darkness into light as shown both by the action maxima at 660 nm and by photoreversibility (Price and Klein, 1961; R. M. Smillie, personal communication). While much remains to be learned about the course of these changes in plastids, they nevertheless show involvement of P_{fr} in differentiation and organization at the subcellular level where transport between cells is not involved, as it is in control of flowering.

RAPID ACTION OF P_{fr}

A drawback to assessing the nature of the action P_{fr} in many of the displays listed in Table 1 is that development of the display re-

TABLE 1

Some Aspects of Plant Development Subject to Modification by Change in Form of Phytochrome

Flowering	Axis modification
Initiation	Cormel formation
Development	Rhizome formation
Cleiostogamy	Bulbing
Phylloidy	Leaf enlargement
Sex	Stem length
Metabolism	Unfolding of the plumular hook in seedlings
Seed respiration	Leaf abscisson
Respiration in Crassulaceae	Epinasty
Anthocyanin formation	Hair formation
Dormancy induction and release	Gametophyte formation in lower plants
Seeds [germination]	Gemmae (*Marchantia*)
Spores [germination]	Induction of the Conchocelis phase (*Porphyra*)
Terminal buds	Haploid prothallium (*Dryopteris*)
	Plastid changes
	Development of lamella
	Orientation (*Mougeotia*)

FIG. 5. The photoreversible effect of far-red (FR) and red (R) radiation in sequence on the closure of *Mimosa pudica* leaflets. (Fondeville *et al.*, 1966.)

quires days or weeks following the stimulus. An indication of very rapid P_{fr} action is shown by suppression of the flowering response of the Japanese morning glory, *Pharbitus nil.* The reversal of action of red by far-red light is possible only in the first 30 seconds after irradiation with red light (Fredericq, 1964), which implies completion of the action of P_{fr} in 30 seconds even though the display, i.e., the failure to flower, can be assessed only after many days. A much quicker display is control of leaf movement, which for the sensitive plant (*Mimosa pudica*) and several other legumes is evident in 15 minutes (Fondeville *et al.*, 1966; see Fig. 5). Orientation of plastids in an alga (*Mougeotia*) is controlled with equal rapidity

(Haupt, 1959). A response requiring only a few seconds is change in properties of a mung bean root after P_{fr} is established by red light (Tanada, 1968). A segment at the tip of the root when suspended in a solution in a glass beaker adheres to the bottom of the beaker after exposure to red light. It is quickly released following a subsequent exposure to far-red light. The cycle can be repeated many times. The exposure to red light, which establishes P_{fr} and causes adherence, is accompanied by a change in potential between the root and the ambient solution (Jaffe, 1968). Other rapid responses are changes in the level of some enzyme activities in white mustard seedlings (*Brassica alba*), which either increase linearly with time or are inhibited after P_r is changed to P_{fr}. Phenylalanine ammonia lyase and lipoxygenase are illustrative (Karow and Mohr, 1969). Some of these responses and all the potentiated ones are photoreversible.

The leaf movements and the conductance changes are membrane dependent. Control of enzyme production by P_{fr} has been interpreted as operating at the level of gene action because it can be suppressed with some antibiotics (Rissland and Mohr, 1967). It too, however, could depend on changes in cellular compartmentation and hardly subverts evidence for the primary control of P_{fr} being on a membrane.

THE MANNER OF P_{fr} ACTION IN FLOWERING AND OTHER RESPONSES

Several other aspects of light influence on plants lead to further understanding of flowering and other responses to change in the form of phytochrome. While the leaves of *Mimosa pudica* and those of many other legumes close in darkness at moderately low levels of P_{fr}, they remain open in continued sunlight, which maintains P_{fr} at these same levels (Fondeville *et al.*, 1967). Some further photoreaction in sunlight must be preventing the closing action of P_{fr}. The action spectrum for maintaining the open leaflet shows a maximum effectiveness near 720 nm. Many seed are suppressed in germination by long exposure to light, even though their germination is promoted by short exposures to red light. The action spectrum for the suppression is similar to that for keeping *Mimosa* leaves open, i.e., with a maximum near 720 nm (Hendricks *et al.*, 1968).

The flowering of many long-day requiring plants, examples of which are wheat, annual beet, and henbane, also responds to prolonged irradiation with a maximum effectiveness near 720 nm. We have attempted without success to detect a pigment, other than phytochrome, having an absorption maximum in this region. This failure and unsuccessful attempts by us and others to involve photosynthesis in the process have led instead to the tentative conclusion that phytochrome is the effective pigment. Debate is current, however, as to how P might be acting. Our concept is that P_{fr} can act only by association with something else—a gene, a membrane, or some cellular component, which can be called a site or place of action. The light absorption of P_{fr} in the associated form must be maximal near 720 nm instead of 730 nm as for free P_{fr}, and its absorption in the region of 660 nm must be greatly reduced with respect to both P_{fr} and P_r to explain the observed responses. P_{fr} is thought to associate with the site by a thermal reaction but be dissociated from the site by a photochemical act with a maximum effectiveness near 720 nm (Fig. 6). A steady state is thought to be established in continuous light between the photodissociation and thermal recombination. This steady state should depend on the intensity and the spectral distribution of the continuously exciting light, as is in fact observed.

RESPONSES DEPENDENT ON PHOTOSENSITIZED OXIDATIONS

Logical treatment of the photochemical aspects of the subject leads me next to discuss several responses to light that depend on

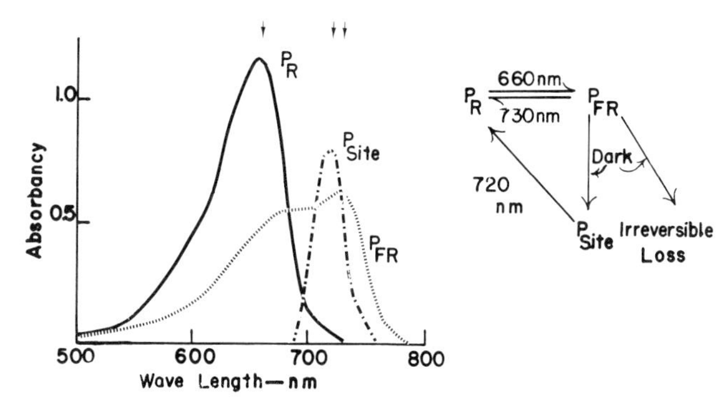

FIG. 6. A possible absorption spectrum of P_{fr} when associated with its site of action compared with absorbancies of P_r and P_{fr}. A partial scheme for light action is shown on the right.

photosensitized oxidations, rather than change of molecular form as in phytochrome. These are carotenoid synthesis by several microbes, lysis in a bacterium and in erythrocytes, and gamete release by several invertebrates. Synthesis of β-carotene, induced by light from the colorless precursor phytoene, has been studied in the bacteria *Myxococcus xanthus* (Burchard and Hendricks, 1969) in the stationary phase of growth and *Mycobacterium marinum* (Batra and Rilling, 1964) and in the fungus *Neurospora crassa* (Zalokar, 1955). The synthesis in each case requires oxygen. Lysis in a bacterium is illustrated by the response to light of *Myxococcus xanthus* in the late stationary phase of development (Burchard *et al.*, 1966). Both this response and hemolysis of erythrocytes (Blum, 1941) require oxygen. Effects of light on gamete release are illustrated by responses of the hydrozoan *Hydractinia echinata* (Jenner *et al.*, 1954) and the solitary ascidians *Ciona intestinalis* and *Molgula manhattensis* (Whittingham, 1967).

Action spectra for carotenoid synthesis in *Myxococcus xanthus* (Fig. 7) indicate that the photoreceptive pigment is protoporphyrin IX, which also is extruded in increasing amounts into the medium

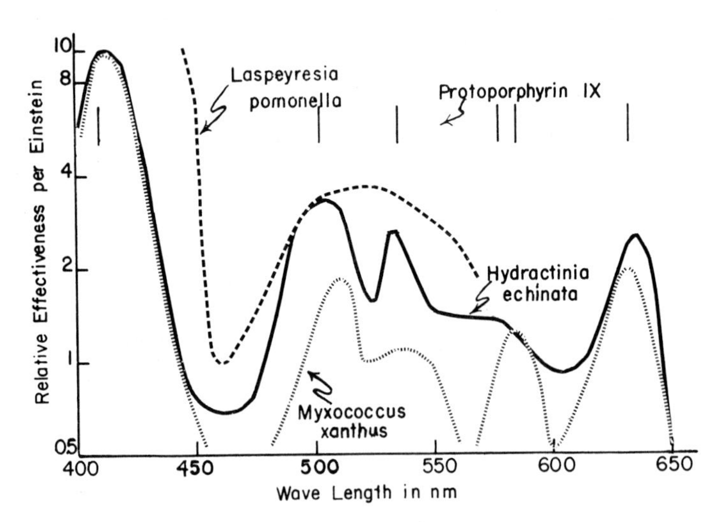

FIG. 7. Action spectra for breaking of diapause of codling moth larvae (*Laspeyresia pomonella*) (Hayes *et al.*, 1969); gamete release by a hydra (Jenner *et al.*, 1954); and carotenoid formation by a bacterium (*Myxococcus xanthus*). (Burchard and Hendricks, 1969.) Positions of absorption maxima of protoporphyrin IX are shown by vertical lines.

as the culture ages. The same pigment causes photolysis of the bacterium in the late stationary phase of growth. A porphyrin, which is probably protoporphyrin IX, is also effective in gamete release in *H. echinata* (Fig. 7). As oocytes develop in darkness, the germinal vesicle enlarges, becomes granular, and moves to the distal wall. Exposure to full sunlight for a microsecond can initiate further development. The first polar body appears in about 45 minutes and often divides before the appearance of the second polar body. The gonophore ruptures between 55 and 60 minutes after the light flash at 25° nm. The process of maturation is initiated solely by the action of light on the oocyte. Less than 10^{-4} erg/oocyte of incident radiation is required at the action maximum (410 nm) for half release. This corresponds to less than 3×10^{-17} mole of product of photoreceptor per oocyte, which is about 10^{-7} *M*. Light-induced spawnings of *Ciona intestinalis* and *Molgula manhattensis* also are probably affected by protoporphyrin IX. In many instances of photosensitization in man, porphyrins are also the probable photoreceptors (Blum, 1941).

Light is effective in the blue spectral region (400–500 nm) for carotenoid synthesis in *N. crassa*. The phase of the circadian rhythm in conidia formation by *N. crassa* var. Timex (Sargent and Briggs, 1967) is also set by light in this region. The photoreceptive pigment is probably a flavin. Riboflavin was present in mycelia of the dark-grown fungus (Zalokar, 1955). Carotenoid synthesis in the strain of *Mycobacterium marinum* studied involves both flavin and porphyrins as photoreceptors. This lack of requirement for a specific photoreceptor is characteristic of photosensitized oxidations (Blum, 1941; Foote, 1968). Thus, in the case of erythrocytes, photolysis can be induced by eosin or other photodynamic dyes added to the suspending medium.

LIGHT ACTION IN CONTROL OF INSECT DIAPAUSE AND ACTIVITY

The function of environmental variables, of which light is one of several, in control of differentiation is best shown by diapause in insects. More basically, in this case the hormonal control of development can be appreciated as the central phenomenon, to which light action and rhythmic change are corollary. Diapause is a dormant or resting stage with negligible or reduced metabolism and suspended growth. Its onset is under hormonal control, which in turn is initiated by the environment. The diapausing form is more resistant to cold, drought, lack of food, or other adverse factors than is the developing insect. General reviews of the extensive literature on diapause and associated hormonal actions are those of Wigglesworth (1954), Lees (1955, 1968), and Engelmann (1968). References dealing with rhythmic displays include Harker (1961, 1964), Adkisson (1966), and Danilevskii (1965).

In a given species diapause occurs at a particular stage, which may be embryonic, larval, pupal, or in the imago. No insect is known to enter diapause before blastoderm formation; a number enter it as adults (imago phase)—the apple-blossom weevil is an example (Danilevskii, 1965). Display of diapause in the embryo stage is well illustrated by the eggs of the silkworm moth, *Bombyx mori*. In those races having two generations in one year, eggs of the second generation are arrested in development at the dumbbell-shaped but unsegmented stage of the embryo. The extensive surgical and endocrine transplant work of Fukuda (1952) and Hasegawa (1952) showed that the onset of diapause in the egg is determined by a hormone from the subesophageal ganglion of the female moth. The presence of this hormone in the hemolymph of the female moth is in turn controlled by a hormone from the neurosecretory cells of the brain.

Diapause in the larvae of the codling moth (*Laspeyresia pomonella*) and many other species, where the individual's own endocrine system is developed, depends on the level of action of a juvenile hormone secreted by the *corpus allatum*. This hormone interacts with other hormones from the prothoracic gland (molting hormone, ecdysone) which in turn is under control of a hormone release from the *corpus cardiacum*. Release of both hormones from the respective glands is under the control of a hormone or hormones from the neurosecretory cells of the brain. The actions of ecdysone and juvenile hormone determine the type of cuticle formed by the insect.

Induction of diapause takes place in short days for the silk worm eggs as well as the larvae of many species. Resumption of growth of the diapausing form of most of these insects requires a period of many days at low temperatures (0–10°). The period at low temperatures apparently leads to the destruction of an inhibitor of the action of the subesophageal hormone. The photoperiodic induction of diapause in these cases and the effectiveness of a period at low temperatures in restoring growth are similar to responses of buds of perennial plants in temperate climates. Dormancy in the buds is induced in the short days of autumn. Overwintering or a period of chilling is required before growth of the bud can resume. In a few insect species, diapause, after induction on short days, can be broken by returning to long days without holding the diapausing form at low temperatures. Such a response is shown by larvae of the codling moth (Hayes *et al.*, 1969) and pupae of the oak silkworm *Antheraea pernyi* (Danilevskii, 1965).

The dependence of diapause on the day or night length, i.e., on photoperiodic conditions, is indicative of some type of timing process as well as of the action of a pigment, or pigments, for photoreception. Some type of rhythmic change within the insect is widely considered to be involved in the timing (Bünning, 1963; Harker, 1961). Circadian or rhythmic changes, however, are better shown in insect activity than in diapause. Regulation of activity of the cockroach was shown by Harker (1964) to be under the control of a secretion from cells of the subesophageal ganglion. Harker transplanted this ganglion from a cockroach having a distinct rhythmic activity into an arhythmic one. Periodic activity was induced in the latter insect.

The variation with time of the activity of a cockroach in darkness can be displaced in time by a light signal (Bruce, 1960). This shift is known as "phasing" of the activity. The extent to which the phasing involves the same photoreaction or reactions as take part in induction or breaking of diapause is an open question at this time.

ACTION SPECTRA OF DIAPAUSE CONTROL

Action spectra of adequate precision for first discussion of the pigments and photoprocesses involved in control of diapause have been measured for the green vetch aphid *Megoura viciae* (Lees, 1966; Fig. 8) and the codling moth *Laspeyresia pomonella* (Hayes *et al.*, 1969; Figs. 6 and 9). Parent viviparae of the green vetch aphid developing on long days bear viviparous parthenogenetic daughters. On long nights (over 9.75 hours) egg-laying females, oviparae, are born, which produce diapausing eggs. Production of oviparae is prevented by 1-hour light interruptions before the 2nd, or after the 6th, hour of 10.5-hour nights. Light is ineffective near the 4th hour. An action spectrum at 1.5–2.5 hours is shown in Fig. 8. The upward arm of the response curve measured at 7.5 hours is shifted 30–50 nm toward longer wavelengths relative to that at 0.5–2.5 hours. It thus extends to 550 nm or beyond.

Action spectra for the codling moth were measured by exposing the larvae in different spectral regions during 6-hour extensions

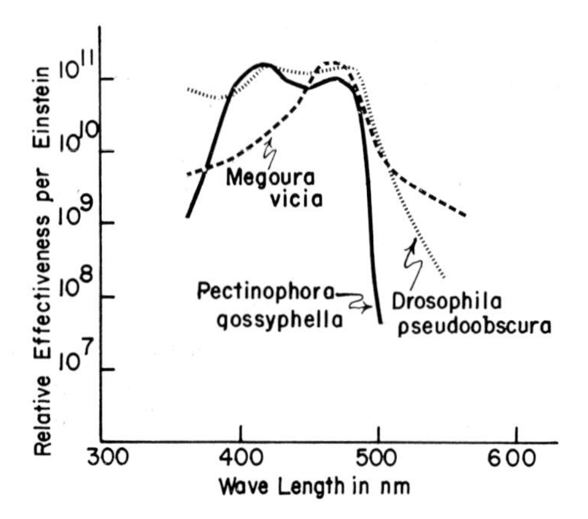

FIG. 8. Relative effectiveness of radiation in the region of 360–560 nm for release of diapause in eggs of the aphid *Megoura viciae* (Lees, 1966); induction of a rhythm in egg hatching of the pink bollworm, *Pectinophora gossyphella* (Bruce and Minis, 1969); and induction of phase shift in adult emergence of *Drosophila pseudoobscura* (Frank and Zimmerman, 1968). The scale is the absolute one for *P. gossyphella* in terms of incident Einstein $^{-1}$ cm^{-2} for 50% emergence.

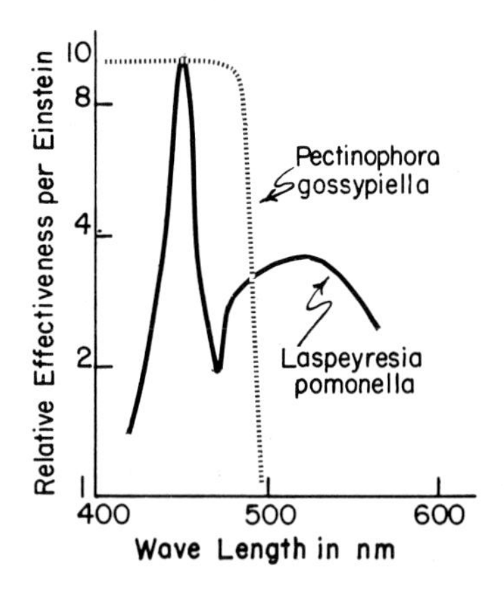

FIG. 9. Relative effectiveness of radiation in the region of 400–560 nm for induction of a rhythm in egg hatching of *Pectinophora gossyphella* (Bruce and Minis, 1969) and diapause release for larvae of the codling moth, *Laspeyresia pomonella* (Hayes *et al.*, 1969.)

of 10-hour days. These spectra too show effectiveness of light in the 500–600 nm region in breaking of diapause (Figs. 7 and 9).

Action spectra for induction of a rhythm in egg hatching of the pink bollworm, *Pectinophora gossypiella*, which can be induced with a single light exposure of low energy, are shown in Figs. 8 and 9 (Bruce and Minis, 1969). Energies as low as 10 erg cm^{-2} in a period of less than 2 minutes were adequate for a threshold response in the region of maximum effectiveness. This is not more than 10-fold greater than required for 50% gamete release by *Hydractinia echinata*. A similar action spectrum is shown for the phase shifts in the rhythmic adult emergence in *Drosophila pseudoobscura* (Frank and Zimmerman, 1969; Fig. 8).

It is evident that the several action spectra are quite different. The effective pigments can hardly be the same even though the action spectra, which are expressed as incident energies required for a given response, are possibly influenced by screening pigments. Thus, the rapid increase in energy for the codling moth at wavelengths below 440 nm could be a result of absorption by the brown coloring material of the larva.

The initial light action in the insect diapause, as a speculation, is considered to be on the membrane permeation of the neurosecretory cells of the brain. The process can be one of photosensitized oxidation much as is involved in gamete release in *Hydractinia echinata*. The exact nature of the release process and the organ or organs involved are still to be discovered, as also are the absorbing pigments. Possible pigments to be considered include flavins, porphyrins, and semiquionones.

SUMMARY

Change in the molecular form of the chromoprotein phytochrome, *P*, by light acts as a control of flowering, seed germination, and many other aspects of plant development and growth. The control probably involves association of the P_{fr} form of *P* with a membrane, resulting in a change of permeability. This change coordinates rhythmic interplay of hormone action with development.

Light also acts through photosensitized oxidations to control some aspects of both plant and animal development. The sensitizing pigments include protoporphyrin IX and possibly some flavins. The actions again involve changes in membrane function, as illustrated by release of a hormone from the neurosecretory cells of the insect brain leading to control of diapause.

REFERENCES

Adkisson, P. L. (1966). Internal clocks and insect diapause. *Science* **154,** 234–241.

Blum, H. F. (1941). "Photodynamic Action and Diseases caused by Light." Reinhold, New York.

BATRA, P. P. and RILLING, H. C. (1964). On the mechanism of photoinduced carotenoid synthesis: aspects of the photoinductive reaction. *Arch. Biochem. Biophys.* **107,** 485–492.

BORTHWICK, H. A., HENDRICKS, S. B., and PARKER, M. W. (1948). Action spectra for photoperiodic control of floral initiation of a long-day plant, Wintex barley (*Hordeum vulgare*). *Botan. Gaz.* **110,** 103–118.

BORTHWICK, H. A., HENDRICKS, S. B., and PARKER, M. W. (1952). The reaction controlling floral initiation. *Proc. Natl. Acad. Sci. U.S.* **38,** 929–934.

BRUCE, V. G. (1960). Environmental entrainment of circadian rhythms. *Cold Spring Harbor Symp. Quant. Biol.* **25,** 29–48.

BRUCE, V. G., and MINIS, D. H. (1969). Circadian clock action spectrum in a photoperiodic moth. *Science* **163,** 583–585.

BÜNNING, E. (1963). "The Physiological Clock." Springer, Heidelberg.

BURCHARD, R. P., and HENDRICKS, S. B. (1969). Action spectrum for carotenogenesis in *Mycococcus xanthus*. *J. Bacteriol.* **97,** 1165–1168.

BURCHARD, R. P., GORDON, S. A., and DWORKIN, M. (1966). Action sprectrum for the photolysis of *Myxococcus xanthus*. *J. Bacteriol.* **91,** 896–897.

DANILEVSKII, A. S. (1965). "Photoperiodism and Seasonal Development of Insects." Oliver & Boyd, Edinburgh.

ENGELMANN, F. (1968). Endocrine control of reproduction in insects. *Ann. Rev. Entomol.* **13,** 1–26.

FONDEVILLE, J. C., BORTHWICK, H. A., and HENDRICKS, S. B. (1966). Leaflet movement of *Mimosa pudica* L. indicative of phytochrome action. *Planta* **69,** 357–364.

FONDEVILLE, J. C., SCHNEIDER, M. J., BORTHWICK, H. A., and HENDRICKS, S. B. (1967). Photocontrol of *Mimosa pudica* L. leaf movement. *Planta* **75,** 228–238.

FOOTE, C. S. (1968). Mechanism of photosensitized oxidation. *Science* **162,** 963–970.

FRANK, K. D., and ZIMMERMAN, W. F. (1968). Action spectrum for phase shifts of a circadian rhythm in Drosophila. *Science* **163,** 688–689.

FREDERICQ. H. (1964). Conditions determining effects of far-red and red irradiations on flowering response of *Pharbitus nil*. *Plant Physiol.* **39,** 812–816.

FUKUDA, S. (1952). Function of the pupal brain and suboesophageal ganglion in the production of diapause and non-diapause eggs in the silkworm. *Annotationes Zool. Japon.* **25,** 149–155.

HARKER, J. E. (1961). Diurnal rhythms. *Ann. Rev. Entomol.* **6,** 131–146.

HARKER, J. E. (1964). "The Physiology of Diurnal Rhythms." Cambridge Univ. Press, London and New York.

HASEGAWA, K. (1952). Studies on voltinisin in the silkworm, *Bombyx mori* L. with special reference to the organs concerning determination of voltinism. *J. Fac. Agr. Tottori Univ.* **1,** 83–124.

HAUPT, W. (1959). The chloroplast turning of *Mougeotia*. *Planta* **53,** 484–502.

HAYES, D. K., ADLER, V. E., SULLIVAN, W. M., SCHECHTER, M. S., NORRIS, K. H., and HOWELL, F. (1969). The action spectra for breaking diapause in the codling moth *Laspeyresia pomonella* L. and the oak silkworm *Antheraea pernyi* Guer. *Proc. Natl. Acad. Sci. U.S.* **64,** in press.

HENDRICKS, S. B., and BORTHWICK, H. A. (1965). The physiological functions of phytochrome. *In* "Chemistry and Biochemistry of Plant Pigments" (T. W. Goodwin, ed.), pp. 405–436. Academic Press, New York.

HENDRICKS, S. B., TOOLE, V. K., and BORTHWICK, H. A. (1968). Opposing action of light in seed germination of *Poa pratensis* and *Amaranthus arenicola*. *Plant Physiol.* **42,** 2023–2028.

JAFFE, M. J. (1968). Phytochrome-mediated biolectric potentials in mung bean seedlings. *Science* **162,** 1016–1017.

JENNER, C. A., PARIS, O. H., HENDRICKS, S. B., and BORTHWICK, H. A. (1954). The action spectrum for gamete release from *Hydractinia echinata*. *Proc. 1st Intern. Photobiol. Congr. Amsterdam*, (P. B. Rotter, M. Sangster, and J. A. J. Stolwijk, eds.), pp. 75–77. Veenman & Zonen, Wageningen.

KAROW, H., and MOHR, H. (1969). Phytochrome-mediated repression of enzyme increase (lipoxidase, E.C. 1.13.1.13) in mustard seedlings (*Sinapis alba* L.). *Naturwissenschaften* in press.

KASPERBAUER, M. J., BORTHWICK, H. A., and HENDRICKS, S. B. (1963). Inhibition of flowering of *Chenopodium rubrum* by prolonged far-red radiation. *Botan. Gaz.* **124,** 444–451.

LEES, A. D. (1955). "The Physiology of Diapause in Arthropods." Cambridge Univ. Press, London and New York.

LEES, A. D. (1966). Photoperiodic timing mechanism in insects. *Nature* **210,** 986–989.

LEES, A. D. (1968). Photoperiodism in insects. *In* "Photophysiology" (A. C. Giese, ed.), Vol. IV. Academic Press, New York.

MOHR, H. (1965). Light regulation of fern gametophyte development. *Ber. Deut. Botan. Ges.* **78,** 54–68.

PRICE, L., and KLEIN, W. H. (1961). Red, far-red response and chlorophyll synthesis. *Plant Physiol.* **36,** 733–735.

RISSLAND, I., and MOHR, H. (1967). Phytochrome-mediated enzyme formation (phenylalanine deaminase) as a rapid process. *Planta* **77,** 239–249.

SARGENT, M. L., and BRIGGS, W. R. (1967). The effects of light on a circadian rhythm of conidiation in *Neurospora*. *Plant Physiol.* **42,** 1504–1510.

TANADA, T. (1968). A rapid photoreversible response of barley root tips in the presence of 3-indoleacetic acid. *Proc. Natl. Acad. Sci. U.S.* **59,** 376–380.

WHITTINGHAM, D. G. (1967). Light induction of shedding of gametes in *Ciona intestinalis* and *Molgula manhattensis*. *Biol. Bull.* **132,** 292–298.

WIGGLESWORTH, V. B. (1954). "The Physiology of Insect Metamorphosis." Cambridge Univ. Press, London and New York.

ZALOKAR, M. (1955). Biosynthesis of carotenoids in *Neurospora*. Action spectrum of photoactivation. *Arch. Biochem. Biophys.* **56,** 318–325.

DEVELOPMENTAL BIOLOGY SUPPLEMENT 3, 244–250 (1969)

Communication in Development: a Postscript

ANTON LANG

MSU/AEC Plant Research Laboratory, Michigan State University, East Lansing, Michigan

I

That communication is a crucial requirement in development will hardly be disputed. Development proceeds in an orderly manner, according to a very precise temporal and spatial plan, nor can it proceed otherwise. Thus, in any developing organism, be it unicellular or consisting of a great number of cells arranged in tissues and organs, there must be continuous communication, including transmission of signals from the environment wherever environmental conditions initiate or control development.

The question that can be raised is whether developmental communication is *a* phenomenon, that is, whether there are some essential principles underlying the various kinds of communication which can be found in development, especially of the more complex, multicellular organisms, or at least whether we can ask questions that pertain to communication phenomena in development as a whole. There is no doubt that it would have been easy to organize a symposium of the same length as this one, but limited to any one of its topics—communication within the cell, communication between cells, hormonal communication, or communication from the environment—or even selecting one particular aspect of one of these topics. For example, Adler and Fishman mention in their paper that the interactions macrophage → lymphoid cell on which they have dwelt are paralleled by data showing the reverse interaction, the release by sensitized lymphoid cells of one or more factors profoundly affecting the macrophages. These two-way interactions, the precise nature of the various messengers, and a number of other aspects of communication in the immunoresponse could easily form the topic of another two and a half days' symposium (and, symposia getting more and more popular, undoubtedly have and will). But is there much point in looking at developmental communication as a whole, in its various phenomenological forms and at all the different levels of organization, as we have tried to do in our symposium?

When considering topics for this symposium I worried about this question, and consulted a number of colleagues, all equally knowledgeable men. Their opinion was split about 50–50, and in the end I decided to proceed. Now, after the meeting and conceding that I may be prejudiced, I feel that the gamble has paid off, and am making use of this space to summarize some results and ideas, as well as to point out some problems which have emerged in the course of the symposium and which in my opinion concern developmental communication at large.

II

The most general trend, which became quite evident in the course of the meeting, was that away from work on developmental problems based on a direct, straightforward application of the Jacob-Monod model. It is becoming quite clear, under the influence of an increasing body of experimental evidence, that the very first reactions of the developmental effectors which are especially characteristic of higher organisms—hormones, "photomorphogenetic" pigments—do not occur directly at the genes. The evidence is both positive, i.e., direct proof of a reaction between the effector and some cellular entity other than DNA, gene, or chromosome; and negative, the kinetics of the response to the effector being incompatible with activation of genes and consequent formation of mRNA and proteins. The prime candidates for the sites of those very first reactions are receptor proteins and membranes. The case for the former is made by Jensen and his co-workers, in their beautiful work on "estrophiles" in the uterus and other target tissues for estrogens. Jaffe has been the most vigorous spokesman for membranes, but the role of membranes as the sites of the first reactions which initiate the developmental communication sequence, whether the initiating agent is a hormone, or light, or something else, became also quite clear in the papers by Williams, Ray, and Hendricks. The effects of ecdysone on the permeability and hence the properties of membranes, mentioned by Williams, are especially noteworthy, as ecdysone has perhaps been the prime example for direct gene activation by a hormone.

Receptor proteins and membranes need by no means be mutually exclusive as sites for the first interaction with the developmental effectors. If estrogen is administered to the uterus *in vitro* no reaction between hormone and receptor proteins takes place. But if the uterus is excised after a brief period of estrogen application *in vitro* the reaction proceeds as *in situ*. This may be an indication that the initial interaction occurs at the membrane level. A common feature of the first interaction between effector and cell, whether with a receptor protein or in a membrane, seems to be that it is of a physical rather than a chemical nature, not involving covalent binding. Allosteric changes in the receptor as the next change or communication step are a possibility that merits close attention. Jaffe has presented evidence for the fascinating possibility that membrane changes may drive, and may perhaps in turn respond to, developmental currents which may result in a polarized distribution of charged cell components by self-electrophoresis. This is in line with the fact that many if not most surface changes in cells are transportive rather than synthetic (e.g., ion movement, auxin transport), and may be the first step toward a more general understanding of the significance of electric potentials and resistances found across or along various cells and tissues.

III

When speaking of the trend away from developmental work based on a *direct*, *straightforward* application of the Jacob-Monod model, I tried to choose words carefully, in order not to create the impression that this earlier work was beside the point, or that further work of this kind has no *raison d'être*. The Jacob-Monod model was the first comprehensive and detailed scheme to explain regulation of gene activity. If developmental biologists accept the idea that regulation of gene activity is an essential aspect of development, they have no choice but to test whether and to what extent that first specific scheme for such a regulation is able to explain development in general, including such problems as the action of hormones and of photomorphogenetic pigments. It is also clear, and to a large extent thanks to such Jacob-Monod based work, that the vast majority of developmental changes in higher organisms, even such relatively simple ones as increase in size of plant cells, do involve the synthesis of new or at least of more RNA and proteins. It remains an entirely legitimate question to ask what these RNA and proteins are and how, exactly, their synthesis is regulated. As clearly illustrated by Gurdon, communication from the cytoplasm to the nucleus remains no less legitimate and important an area of developmental research. It may turn out that the "complications" introduced by the early reactions of effectors with receptor proteins or membranes are not very large. Jensen and co-workers hypothesize that the estrogen-receptor entity removes some component in the nucleus that blocks the activity of certain genes. We may on the other hand find, at least in other cases, that the situation is a good deal more complex, with many more intervening steps. But we have in any event gained a broader and more perspective conceptual framework in which to see the problems and to plan experiments for their solution.

IV

Two other developments which have also been brought out in the symposium add to the broadening of our concepts and will help us in bringing the problem of developmental communication into sharper focus. The first is evident in Maaløe's model to explain the adjustment of the bacterial cell to growth in different environments. One of the assumptions on which this model is resting is that the fraction of the mRNA molecules which codes for ribosomal protein is governed by the *overall pattern* of gene repressions and inductions, which in turn constitutes the cell's response to its environment. The classical Jacob-Monod model would postulate an operon of rRNA-protein genes controlled by a specific effector the level of which would in turn somehow be controlled by the growth rate, although growth rate can be varied in many different ways. With its alternative assumption, the Maaløe model provides a more plausible and more economic explanation for the fact that the number of ribosomes in the bacterial cell is proportional to the growth rate (and their synthesis thus to the square of this rate), and that the ribosomal proteins and hence the corresponding mRNA seem always to be produced in amounts neatly matching those of the rRNA. Maaløe's model should stimulate similar considerations for the cells of higher organisms, particularly since it is already known that the number of ribosomes in these cells may be closely correlated with the rate of growth or development. But the most important feature of Maaløe's model, to my mind, is that it goes an important step beyond the basic Jacob-Monod model, and does so in that material in which the latter model was developed.

The second development, or rather viewpoint, was most clearly brought out in the discussion following Filner's paper, by R. T. Schimke and others. It is that we ought to view the bacterial and the eucaryotic cell both as products of the *evolution* of communication systems, and therefore if we want to make projections from the former to the latter go by route B, rather than route A, which we have been mostly using:

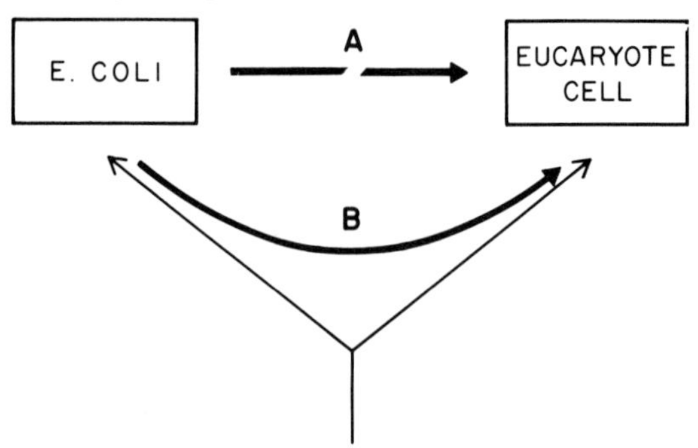

If we do this, then many observations which we now tend to view as bothersome exceptions to the *E. coli* based dogma, and which we therefore may try to accommodate into that dogma, sometimes making the latter somewhat of a Procrustean bed, assume a quite different image: they may be parts—perhaps parts still widely disconnected—of a communication and regulation system characteristic of, and limited to, the eucaryote cell, arisen during its evolutionary history. This may be true of several findings described in the course of this symposium, such as the effects of cytoplasmic components on nuclear activity accompanied by dispersion of chromosome material, as reported by Gurdon; or the RNA-antigen complexes described by Adler and Fishman. Viewed in this manner, some ideas proposed during the symposium also gain considerably in interest: the "commander" proposed by Filner—a mechanism to enable the cell of the higher organism to adjust its behavior to the multicellular way of life; the importance for cells of higher organisms to be able not only to synthesize but also to inactivate enzymes, mentioned by Filner and further stressed in the discussion after his talk; the suggestion to distinguish between true regulatory genes that regulate through the cytoplasm the synthesis of specific RNA species, and structural genes either of which may affect protein (enzyme) levels, made by Schimke, also following Filner's paper.

V

The importance and usefulness—rather, the necessity—of viewing developmental communication in an evolutionary or historical perspective became apparent also in another part of the symposium, the papers by Pattee and Varner. Varner's review was concerned with experimental approaches done under conditions thought to approximate those of prebiotic times; it showed to how large an extent it has been possible to produce in this manner biologically important molecules and obtain biologically important activities, including communication between polypeptides and nucleotides. Pattee—whose paper, at the risk of offending the other speakers, I found the most thought-provoking of the program—asked how to distinguish genuine *communication* between molecules from the normal physical *interactions* between molecules, and started the discussion from the premise that it is not the properties of a molecule which make it a carrier of communication—a messenger—but the constraints of the "language" in which it has to operate. He also made the important point that a messenger should carry the message but not be involved in its expression or use, and I was asking myself whether this may not account for the notable failure of studies—some of them very extensive—on the structure-activity relationships of hormones to contribute anything important to our understanding of hormone action. However, Pattee's other point, that a molecule becomes a message only through the integrated system of constraints of the language, is of much more general importance. In the evolution of all languages we find that the symbols themselves tend to become simple and clear while the structure of the language, which gives meaning to the symbols, becomes richer and more complex, and we may expect the same of the molecular languages which are used for developmental communication and other needs of the living cell. While the details may be complex we may expect to find some principles of molecular languages, just as we have them in our own, higher languages. There must be a distinction between the genetic and phenotypic functions, between construction and description. The basic types of building blocks should be small, but the syntactical rules of the language must be rich and varied enough to allow unlimited description and construction. When we view the problem in this fashion we cannot escape the uncomfortable feeling that much of our work on developmental communication and on development in general has been concerned with the "symbols" and has neglected the "language."

The normal course of evolution of languages—our human languages and presumably the molecular ones, too—poses one more specific but interesting question. The two most fundamental properties of living matter are enzymatic activity and self-replication. Today, they are strictly separated; the proteins have the one property and the nucleic acids the second. As replication is essential to maintain life we may tend to postulate that nucleic acids must have arisen first, or at least simultaneously with proteins. But as Pattee pointed out in the discussion following his paper, it is not necessary to assume that replication of nucleic acids was the first case of replication in evolution. Although the double alphabet is universal today, it is not necessary to envisage replication. In fact, the more complex parts of a communication system arise probably later than the simpler ones, just as magnetic tapes were introduced into computers after the simple switch.

In some connection with this question, E. Margoliash made an interesting, although as he himself was ready to admit, perhaps a bit far-out, suggestion. It is based on his and his associates' technique of developing statistical phylogenetic trees from the amino acid sequences of a sufficient number of homologous proteins, which also permits to obtain an approximation of the amino acid sequence of the ancestral form at the topmost apex of such a statistical phylogenetic tree. With enough information for many sets of homologous proteins this may be done independently for each, and having, as it were, wiped out the changes which have occurred since the ancestral forms existed, it might be possible to show that these ancestral forms for the different sets of proteins themselves are homologous. Hence, one could obtain a statistical phylogenetic tree using the ancestral forms, leading to a second-order original ancestral form which would be an ancestral form for all the various sets of homologous proteins considered. This sequence is likely to be very near the structure of the original duplicating polypeptide. An essentially similar procedure should be in principle possible with nucleotide sequences in tRNAs and rRNAs from numerous species. This should in turn lead to a primordial nucleic acid sequence which should bear a distinct resemblance to the primordial polypeptide sequence derived from the protein analysis. Considering the rapid advances in automation of sequence analyses, such a test or approximation of the actual history of past changes in macromolecular structure seems by no means out of reach.

ACKNOWLEDGMENTS

I want to thank several participants in the symposium, especially Drs. Margoliash and Schimke, for helping me to reconstruct some of their discussion remarks. However, the manner in which these have been used in this postscript is entirely my responsibility.

AUTHOR INDEX

Numbers in italics indicate the pages on which the complete references are listed.

H

I

J

N

SUBJECT INDEX

Inclusive pages (1–3) = comprehensive treatment; single consecutive pages (1, 2, 3) = single references to subject; italics (*1*, *2*) = major references including definitions, etc.

D

E

INFORMATION FOR AUTHORS

Developmental Biology will publish articles bearing on problems of development in the broadest sense; it will contain papers dealing with embryonic and postembryonic development, growth, regeneration, and tissue repair, of both plants and animals. The journal will serve as meeting ground for botanical and zoological approaches, drawing on concepts and techniques of a wide range of disciplines, e.g., biochemistry, biophysics, cytology, embryology, experimental morphology, genetics, immunology, microbiology, and pathology. Whatever the organism or the method of study used, the principal criterion of acceptability will be the degree of focus on developmental problems. Articles based on the incidental use of developing systems for other purposes will not be accepted.

Address for Submitting Manuscripts: All manuscripts and all inquiries regarding editorial policy or the preparation of papers, should be sent to: DEVELOPMENTAL BIOLOGY, Box G, Brown University, Providence, Rhode Island 02912, U.S.A.

Manuscripts may be submitted in English, French, or German. They should be concise and consistent in style, spelling, and use of abbreviations. They must be typed double-spaced on one side of numbered pages; corrections in the typescript should be printed in ink. Two copies, the original and one carbon, should be submitted. Figure legends, footnotes, and acknowledgments should also be typed on separate pages. A summary not exceeding 500 words should be included at the end of the manuscript. MANUSCRIPTS SUBMITTED IN FRENCH OR GERMAN MUST INCLUDE AN ENGLISH SUMMARY.

Tables should be typed on separate sheets and the place(s) where they are to be inserted in the text marked by the author. Tables should be numbered with Arabic numerals; a brief title should be typed above the table.

Figures. All illustrations should be submitted in original form; duplicate copies should be submitted for editorial use. Line drawings should be prepared on white drawing or tracing paper, or on blue-lined graph paper. The illustrations should be numbered consecutively in order of their mention in the text (line drawings and half-tones or plates should be included in the same sequence). Each illustration should be identified on the back or in a margin with the name of the author(s) and the figure number. The approximate place(s) where figures are to be inserted should be indicated by the author.

A maximum of two pages of half-tone illustrations will be allowed per article. Authors will be charged extra for half-tone illustrations in excess of this amount.

Color plates are published at the author's expense. Specific inquiries should be directed to the Editor-in-Chief.

Literature references in the text should be in one of the following forms: Doe, 1925; Doe *et al.*, 1905; Doe, 1920, p. 250 (for references to a specific page). Suffixes *a*, *b*, etc., should be used following the date to distinguish two or more works by the same author(s) in the same year, e.g., Doe, 1930a, 1930b. Literature citations in the bibliography should be arranged alphabetically according to the surname of the author. Journal abbreviations should be in accord with Chemical Abstracts' *ACCESS, Key to the Source Literature of the Chemical Sciences* (1969 Edition), as in the following examples:

Astbury, W. T., Beighton, E., and Weibull, C. (1955). The structure of bacterial flagella. *Symp. Soc. Exptl. Biol.* **9,** 282–305.

Flexner, L. B. (1950). The cytological, biochemical, and physiological differentiation of the neuroblast. *In* "Genetic Neurology" (P. Weiss, ed.), pp. 194–198. Univ. of Chicago Press, Chicago, Illinois.

Proofs: Galley proofs will be sent to the authors with reprint order forms. Fifty reprints of each article are granted free of charge.